Land Administration

Spatial Information Systems and Geostatics Series

Land Administration

PETER F. DALE

AND

JOHN D. MCLAUGHLIN

OXFORD

UNIVERSITY PRESS

OXFORD
UNIVERSITY PRESS

Great Clarendon Street, Oxford OX2 6DP

Oxford University Press is a department of the University of Oxford.
It furthers the University's objective of excellence in research, scholarship,
and education by publishing worldwide in

Oxford New York
Athens Auckland Bangkok Bogotá Buenos Aires Calcutta
Cape Town Chennai Dar es Salaam Delhi Florence Hong Kong Istanbul
Karachi Kuala Lumpur Madrid Melbourne Mexico City Mumbai
Nairobi Paris São Paulo Singapore Taipei Tokyo Toronto Warsaw

and associated companies in Berlin Ibadan

Oxford is a registered trade mark of Oxford University Press
in the UK and certain other countries

Published in the United States
by Oxford University Press Inc., New York

First published 1999

British Library Cataloguing in Publication Data
Data available

Library of Congress Cataloging in Publication Data
Data available

ISBN 0-19-823390-6

1 3 5 7 9 10 8 6 4 2

Typeset by Best-set Typesetter Ltd., Hong Kong
Printed in Great Britain
on acid-free paper by
Biddles Ltd Guildford and King's Lynn

Preface

Land Administration is a successor to our book on *Land Information Management* published by the Oxford University Press in 1988. It builds on the earlier text, developing a number of ideas embedded in that work but placing them in the context of contemporary land resource management. The present work is intended for developing countries, countries in economic transition from a command to a market driven economy, and developed countries reviewing their land administration systems in the light of modern attitudes and technologies.

Like its predecessor, this book has been written for students and administrators. It focuses on the administrative environment, only discussing technical matters to a depth appropriate for managers. There is of course a danger of taking too broad an approach and hence moving into too shallow waters. There is however need to examine the management of land and property resources from a broad perspective rather than to deal with the tenure, value, and use of land in isolation. All these components of land resource management are interrelated and are affected by information technology and other factors in the environment.

In the intervening decade since the publication of our earlier book, there has been an unprecedented revival of interest in land tenure reform and associated land management activities. This has largely been brought about by the end of communism in Eastern Europe and by the break-up of the former Soviet Union. It has a parallel with the processes of decolonization in the 1960s when Britain and France responded to the so-called winds of change that were then blowing through much of Africa, resulting in major shifts in the relationship between people and their land.

The transition process today is however markedly different from that of thirty years ago, in part because of the speed with which political power was handed over and in part because of the current objectives of land reform. In the land tenure reforms of the 1960s, the main objective was to create stability and to provide security of tenure in order that farmers in particular could remain on their land without fear of eviction. Today, the objectives are in part to meet the call for social justice, and in part to stimulate economic development. In particular, those responsible for providing financial assistance to countries in economic transition have been concerned primarily with the creation of land markets.

The world is also witnessing a massive migration from the countryside to urban areas. In 1970 an estimated two-thirds of the world's population lived in rural communities; today the figure is about half. By about AD 2030 two-thirds of the world's population are likely to be living in towns and cities and 80 per cent are likely to be living within 50 kilometres of the coastline. In the words of the Preamble to the Habitat Agenda that resulted from the Habitat II Conference in 1996:

> The most serious problems confronting cities and towns and their inhabitants include inadequate financial resources, lack of employment opportunities, spreading homelessness and expansion of squatter settlements, increased poverty and a widening gap between rich and poor, growing insecurity and rising crime rates, inadequate and deteriorating building stock, services and infrastructure, lack of health and educational facilities, improper land use, insecure land tenure, rising traffic congestion, increasing pollution, lack of green spaces, inadequate water supply and sanitation, unco-ordinated urban development and an increasing vulnerability to disaster. (UNCHS 1996)

Access to land and security for credit underpin the solutions to most of these problems. There is a need for all States to ensure that good land administration mechanisms are in place. This book examines why and how this may be done. In particular, it

1. provides an introduction to the role that property plays in social and economic development;
2. focuses more specifically on the role and importance of property infrastructure (and especially on the significance of land administration functions);
3. briefly reviews the nature and function of land administration systems;
4. identifies some of the more significant trends and issues associated with modern land administration;
5. highlights selected trends and issues in detail (such as the concepts of re-engineering, technology developments, economic issues, etc.); and
6. examines some recent case studies in the land administration field.

Peter Dale
University College, London
John McLaughlin
University of New Brunswick

December 1998

References

Dale, P. F., and McLaughlin, J. D. (1988), *Land Information Management* (Oxford: Clarendon Press).

UNCHS (1996), *Habitat Agenda and Istanbul Declaration* (Nairobi: United Nations Centre for Human Settlement).

Contents

List of Figures

List of Tables

1

Introduction and Overview

1.1 Land and Society

1948 saw the Universal Declaration of Human Rights. All governments that signed the Declaration formally recognized that human beings have a right to adequate shelter as one component of their right to an adequate standard of living. In 1996, at the Habitat II Conference, many countries committed themselves to 'Promoting optimal use of productive land in urban and rural areas and protecting fragile ecosystems and environmentally vulnerable areas from the negative impacts of human settlements, inter alia, through developing and supporting the implementation of improved land management practises' (UNCHS 1996).

Access to land and security for credit are vital components of sustainable development and good land management practice; every State needs to ensure that efficient and effective land administration mechanisms are in place. Land is a physical thing that encompasses the surface of the earth and all things attached to it both above and below. It is also an abstract thing that is manifest as a set of rights to its use with a value that can be traded even though the physical object cannot be moved. The term 'land administration' is used here to refer to those public sector activities required to support the alienation, development, use, valuation, and transfer of land.

The relationship between human beings and the land is of fundamental importance in every society and is evident in the form of property rights. This relationship has evolved in many different ways, from full state control, through communal forms of tenure, to the individual property rights. In the land reform programmes of the 1960s, especially in Africa, new land tenure arrangements such as the granting of individual freehold rights were introduced, often contrary to traditional custom and practice. In the reforms of today, especially in east and central Europe, the land restitution programmes have reprivatized land previously owned by individuals. They have also granted first-time rights of private ownership to what was formerly state-owned land, on the basis of the traditions of the past, thus reinforcing old land tenure arrangements.

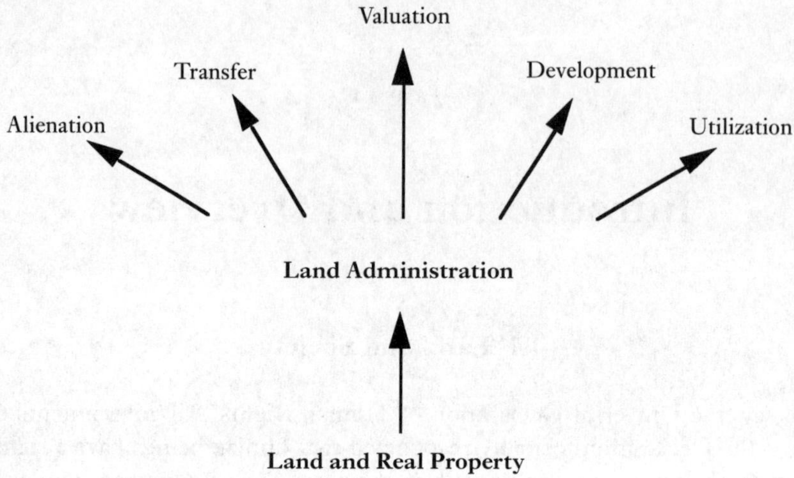

FIG. 1.1 Land administration

1.2 The Importance of Land and Property

The careful management of land and property is fundamental to economic development and the sustainability of the environment, fostering of good governance, and the protection of civil societies. Arguments for the importance of good land administration include:

- Assurance. The more clearly a person's property rights can be defined, the more easily that person will be able to defend those rights against the claims of others (Johnson 1972).
- Social stability. Accurate public records prevent some disputes from arising, and help to resolve others more quickly since property ownership and boundaries can be readily ascertained (Simpson 1976).
- Credit. Accurate public records reduce uncertainty in information by allowing creditors to determine more easily if a potential borrower has the right to transfer property offered as collateral (Feder and Feeny 1991).
- Improvements to the land. Enhanced assurance of property rights increases the certainty that a person will be able to benefit when making investments in buildings, equipment, or infrastructural improvements including land conservation measures. Improved access to credit provides financial resources for improving the land (Feder et al. 1988).
- Productivity. The factors described above combine to ensure that land goes to its highest and best use. As a result, for example, commercial

agriculture is adopted by innovative and energetic farmers who are able to acquire more land; bad farmers end up holding less land (Swynnerton 1955).

- Liquidity. When property rights are formal, assets can be exchanged rapidly on a massive scale and at low cost. In developing countries, most property rights are held informally and so cannot enter the formal market place as negotiable assets (de Soto 1993).
- Labour mobility. Opening up land markets facilitates geographical mobility in the work force; people are able to move from their land in order to benefit from social and economic opportunities offered elsewhere (OECD 1986). The ability to defend rights legally rather than physically allows people to leave their land to take advantage of short-term opportunities.
- Land and property values. Improvements to the land and improved production are reflected in higher land values. Increased labour mobility and liquidity allow values to be defined by the market rather than arbitrarily. Clearly defined property rights reduce the costs of discovering who holds rights to any area of land and the scope of those rights. Lowering the cost of finding information about land and property increases the incentive for potential buyers or lessees to make a higher offer for the land or property (Johnson 1972).
- Land and property taxation. The increased value of the land with clearly identified owners makes the collection of land and property taxes feasible (Dorner 1991).
- Public services. Increased revenues from taxes, together with better information about the land, enable governments and utilities to provide more effective services and utilities (Cymet 1992). Infrastructure such as street lighting, paved roads, and water and sewerage in turn raise property values.
- Resource management. Increased information about the land and property rights allows public agencies and private corporations to plan the management of resources more effectively and enables governments to enforce environmental and other regulations. It acts as a 'meta institution', that is as an enabling mechanism that allows other functions (such as the building of power, communication, and water systems) to be undertaken more effectively (McLaughlin and Nichols 1989).
- Social Development. The role of property in social development is increasingly viewed as being of even more long term importance than its economic contributions (Putnam 1992).

There are however counter-arguments put forward that detract from the importance of land and property. These include:

- Security. Indigenous land and property systems often provide greater security of tenure than previously recognized (Palmer 1996). Claims of subordinate right holders to limited or common access tend to be neglected (Attwood 1990) and tensions can remain between de jure and de facto rights to land, leading to an increase in the number of disputes (Haugerud 1989).
- Incentives. The rights to agricultural land allocated in indigenous property systems need not constrain incentives to invest in improvements (Binswanger and Deininger 1994).
- Credit. Credit is not necessarily related to formal property rights; also, credit institutions simply may not exist for the poor (Migot-Adholla et al. 1991). Where available, political influence may be more important than secure titles (Shipton and Goheen 1992) and credit obtained for agricultural purposes (using the land as collateral) is often diverted for other purposes (Haugerud 1989).
- Optimal land use. Land may already be used for its highest and best purpose, hence consolidation following the formalization of property rights may not lead to more effective farming practices; rather the land may be accumulated for speculation or for future subsistence and security for heirs (Haugerud 1989).
- Costs. The costs of building and maintaining the property infrastructure can be considerable; devoting scarce resources to these activities, especially in very low-income countries, will prevent their use in what may be more important areas (Attwood 1990).

Whereas the good management of land and property is an essential component of economic and social development, land administration must not be treated in isolation from other activities (such as banking and the provision of credit). Land reform in particular is a complex process in which other institutions and functions may be more important than property at any given stage in the development cycle. Good land administration is necessary but not sufficient to ensure that development is beneficial and sustainable.

1.3 Land and Economics

Land, together with its associated buildings and construction, is one of the most important financial assets in any country. Every investment is in some way or another dependent on land and property. Without land no shop or factory can be built, no road or railway constructed, there can be no schools or hospitals, and there can be no government or private sector buildings. Without the security of title to land or buildings it is difficult to obtain investment funds and venture capital. Poor land administration is an impediment to the growth of an economy—banks for example are hesitant to meet the needs for financing without security of title, because of the higher costs and more significant risks. Legal security of land tenure facilitates mortgage-based investment financing for small and middle-scale businesses and underpins the physical infrastructure of almost all commercial operations.

Good land administration contributes to economic development in a number of ways. It provides security to investors and permits real estate to be traded in the market place. It also allows governments to raise taxes on the basis of the value of land and property, either at the time of land transfer (in the form of Stamp Duty or a transfer tax) or directly and annually on the estimated worth of the land or property. This is tantamount to a tax on wealth and has the further advantage that unlike personal income, buildings and the land on which they stand cannot easily be hidden from the tax collector. With good land administration, land and property taxes are easier to administer and can lead to the collection of substantial revenue.

Land and property are financial assets in their own rights and can attract significant amounts of inward investment. A study by the Economist Intelligence Unit into global direct investment noted the importance of real estate and the property sector in foreign direct investment (FDI). It suggested that:

> While there is no doubt as to the importance of FDI in the world economy, the importance of real estate in FDI is difficult to gauge . . . Statistics on foreign direct investment flows into real estate are not standardized. Sectoral breakdowns of FDI flows largely ignore the property market, as most statistical agencies regard real estate as a special case . . . However, a rough estimate on the basis of existing data would imply that the real estate component of FDI could be anywhere between 5–20% of the total. Either way the real estate sector is significant: at the low end of the estimate it attracts more FDI each year than the whole of Eastern Europe and the former USSR, at the high end it attracts nearly double the FDI of China (EIU 1997).

One of the conclusions of the report was that:

> There is a danger of a mismatch between a liberalized global economy with strong free capital flows and a heavily restricted real estate sector burdened by country-specific restrictions. The result will be barriers to FDI and trade. Global harmonization of property markets and standardization of rules and regulations governing real estate are necessary steps to boosting investor confidence and allowing transparency in the FDI regime governing real estate (EIU 1997).

1.4 Land Market Development

One of the key elements in the moves by former communist countries towards a market driven economy has been the privatization of land and buildings and the creation of land markets. In some jurisdictions buildings are regarded as an integral part of the land while in others they are treated as separate entities. In the present context, land will generally be assumed to include buildings. Private ownership of land includes the rights to dispose of it, to use it as collateral for mortgages, to buy and sell it, and to use it subject to certain private and public encumbrances.

The land market operates to encourage mobility such that people can obtain employment in locations beyond the range of their current residential area; it also helps to encourage optimum profit from the land. In many countries land markets are more buoyant in urban than in rural areas. This is in part because those who live in rural communities tend to be more conservative and are more closely attached to the soil. Many farmers for example rate the social value of their land more highly than its market price—their need for land is as much psychological as financial. Urban populations on the other hand tend to be more dynamic than those living in rural areas and are more focused on the economic rather than cultural aspects of landholding. Urban people are also less committed to staying in one place since many are already migrants from other areas of the cities or the country. They are therefore a more mobile workforce and, subject to appropriate economic incentives, are more likely to buy, sell, or rent property than their rural counterparts.

Legal security of ownership with the clear identification of owners and what is owned, mechanisms for efficiently transferring rights in real estate and the provision of effective public access to land information are important prerequisites for developing land markets. Sound land administration within an appropriate institutional regime is essential for these developments. It is not necessarily the function of government to operate these land administration arrangements but it is the responsibility of government to ensure that they are in place and ultimately to be the guardian of the quality of service provided.

Land markets promote the efficient distribution and use of land. Within open markets, owners of land or buildings surplus to their needs can sell them to anyone who can use them more profitably or more efficiently. Thus in theory, over time, market mechanisms should ensure that land moves to its optimum use. The land market (and its associated legal foundation) is thus an instrument of land management policy and as such has a far-reaching economic impact.

It is however a blunt instrument. In the case of rural land reform, and in particular land consolidation, there is a growing awareness that existing markets may be too cumbersome to facilitate change at the pace that is necessary. Concerns for family inheritance and the social rather than the economic value of land may encourage land fragmentation and inhibit land reassembly. Governments therefore need mechanisms to intervene when the market cannot on its own achieve the desired social and economic objectives.

Mechanisms are also needed to allow the State to acquire urban and rural land in the public interest—for instance for infrastructure development such as the construction of new roads, for the provision of State services, and to facilitate land assembly. The compulsory purchase of land must be at prices that are seen to be equitable and procedures for assessing fair compensation and for appeals against such assessments must be put in place. By monitoring the land market through the land administration system, the government can determine the current market prices of land.

1.5 Land Information Management

Developing democracy requires the evolution of responsive governmental institutions. One of the cornerstones of a democracy is an equitable distribution of resources, of which land is fundamental. Uncertain or unobtainable access to land is a cause of disputes in many countries. Uncertainty over the boundaries between neighbours can cause a breakdown in social cohesion while uncertainties over landownership may result either in lengthy and expensive court cases or else in land remaining underutilized. Not only is this a waste of resources but it also reduces land values which in turn reduce the revenues that governments can potentially collect through land related taxes. Such revenue can be used to defray other government expenditure or to fund social reform.

Access to reliable information on the ownership, value, and use of land helps to further social and political objectives. The most common form of land information system is the cadastre. There is no unique form of cadastre for, as Dowson and Sheppard remarked:

> It is impossible to give a definition of a Cadastre which is both terse and comprehensive, but its distinctive character is readily recognized and may be expressed as the marriage of (a) technical record of the parcellation of the land through any given territory, usually represented on plans of suitable scale, with (b) authoritative documentary record, whether of a fiscal or proprietary nature or of the two combined, usually embodied in appropriate associated registers (Dowson and Sheppard 1952).

Land information systems help to facilitate rural land reform, improve urban planning and infrastructure development, and support environmental monitoring. From both a central and local government perspective such systems help to protect land belonging to the state or to local communities—many local councils for example have surprisingly poor information about the real estate assets that they own or for which they are responsible. Similarly the state is a major landowner but the management of its land and buildings is rarely optimal.

All developed societies are moving into the information age with technology that allows easy and open access to data and information, and appropriate mechanisms for recovering the cost of their collection, processing, and distribution. Land related information is an especially valuable commodity in its own right that can be exploited in many different ways, such as through the direct sale of information products and services, or by adding value to service delivery and public decision-making processes.

1.6 Integrated National Land Policies

There are three key attributes of land that every country must manage—its tenure, value, and use. How individuals use their land is every bit as important as how and at what price they can buy and sell it. In many countries the management of use rights is treated separately from the ownership rights yet the value of a property depends both on the certainty of its ownership and the use that can be made of it. In much of east and central Europe, the priority has been to restore individual property rights, a process known as land restitution. This sometimes involved the initial granting of use rights so that agricultural production could continue; ownership rights were restored at a later stage because they were more complex to sort out and required detailed examination of the national archives. All governments recognize the importance of use rights as well as ownership rights but normally choose to manage them differently, often retaining the control of ownership as a central government responsibility while delegating much responsibility for land use control to local government. This is in spite of the fact that ownership affects the use of the land while conversely the use will influence the form and substance of the tenure.

In similar fashion, the manner in which land is taxed can alter the way in which it is used. Systems of land taxation are generally administered by those responsible for revenue collection rather than by those responsible for physical planning. Within a market economy, land taxation can be used as an instrument for physical planning, encouraging profit through certain types of use and loss through others. Yet dialogue between the ministries responsible for Finance and for Physical Planning is relatively uncommon.

The interrelationship between tenure, value, and use was reflected in the Economist Intelligence Unit's report on foreign direct investment already cited when it observed that:

> There are several factors which have an impact on investment flows into property, including restrictions on land ownership; the degree of control foreign investors can exercise over their property; confiscation of property through nationalization,

expropriation or any other means; environmental, health and safety issues; development regulations including planning and zoning rules; variations in taxation on property; the differing role of property consultants in different countries; and international differences in valuation guidelines (EIU 1997).

Rarely have the institutional arrangements been put in place to integrate the ways in which land is managed. Rather, the management of land and property as a national resource has tended to be fragmented.

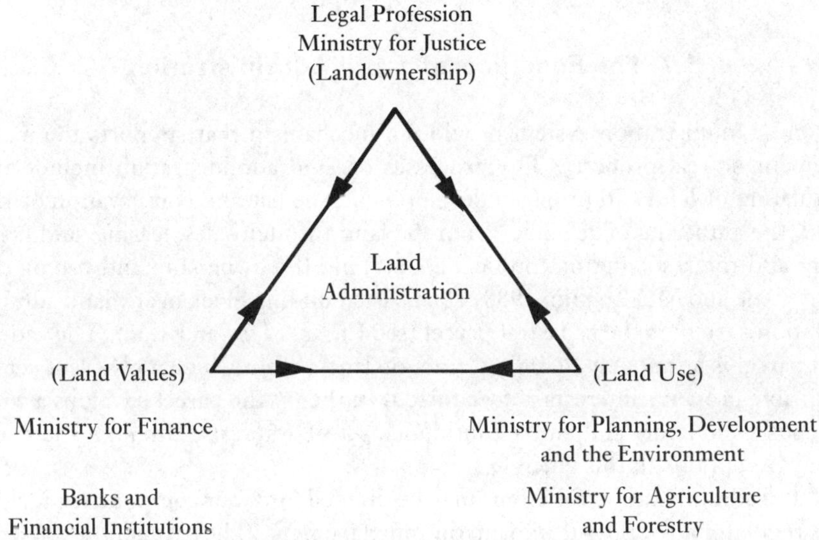

FIG. 1.2 Ownership, value, and use

If land resources are to be used in an optimum fashion then the management of land and its associated resources must operate within an integrated state land policy framework. A holistic approach is necessary if the full benefits are to be achieved. As noted by the United Nations Economic Commission for Europe in its guidelines on land administration:

> The modern cadastre is not primarily concerned with generalized data but rather with detailed information at the individual land parcel level. As such it should service the needs both of the individual and of the community at large. Benefits arise through its application to: asset management; conveyancing; credit security; demographic analysis; development control; emergency planning and management; environmental impact assessment; housing transactions and land market analysis; land and property ownership; land and property taxation; land reform; monitoring statistical data; physical planning; property portfolio management; public communication; site location; site management and protection (UNECE 1996).

The guidelines pointed out that although land records are expensive to compile and to keep up to date, a good land administration system should produce benefits that significantly outweigh the costs. Given the wide range of applications indicated above, some degree of central coordination is necessary. The case for good land administration rests on good commercial grounds as well as upon matters of social justice. The central issue is not whether countries can afford such a system but whether they can afford to live without one.

1.7 The Functions of Land Administration

A land administration system provides a mechanism that supports the management of real property. The processes of land administration include the regulating of land and property development, the use and conservation of the land, the gathering of revenues from the land through sales, leasing, and taxation; and the resolving of conflicts concerning the ownership and use of the land (Dale and McLaughlin 1988). The basic building block in any land administration system is the cadastral parcel (see Figures 1.3*a* and 1.3*b*). The cadastral parcel is a uniquely delimited tract of land within which a coherent set of definable property interests is recognized. In theory the parcel envelops a continuous volume of land and a continuous set of interests, although there are many exceptions to this rule.

Land administration functions may be divided into four components: juridical, regulatory, fiscal, and information management. These functions of land administration are traditionally organized around three sets of agencies responsible for surveying and mapping, land registration, and land valuation. Each of these agencies collects data and makes them available to the public.

The juridical component places greatest emphasis on the holding and registration of rights in land. It comprises a series of processes concerned with the original determination or adjudication of existing land rights, the allocation of land, for example, through original grants from the sovereign power, transfers, prescription, and expropriation. Other processes address the delimitation of parcels by defining the land for which the rights are allocated, demarcating boundaries on the ground, and describing these boundaries graphically, numerically, or in writing. Over time, the rights to land and the delimitation of parcels may give rise to doubts or disputes and there must therefore be further mechanisms for resolving these. Adjudication is the dispute resolution process while registration is the process of making and keeping records of property rights.

The regulatory component is mostly concerned with the development and use of the land. It includes land development and use restrictions imposed

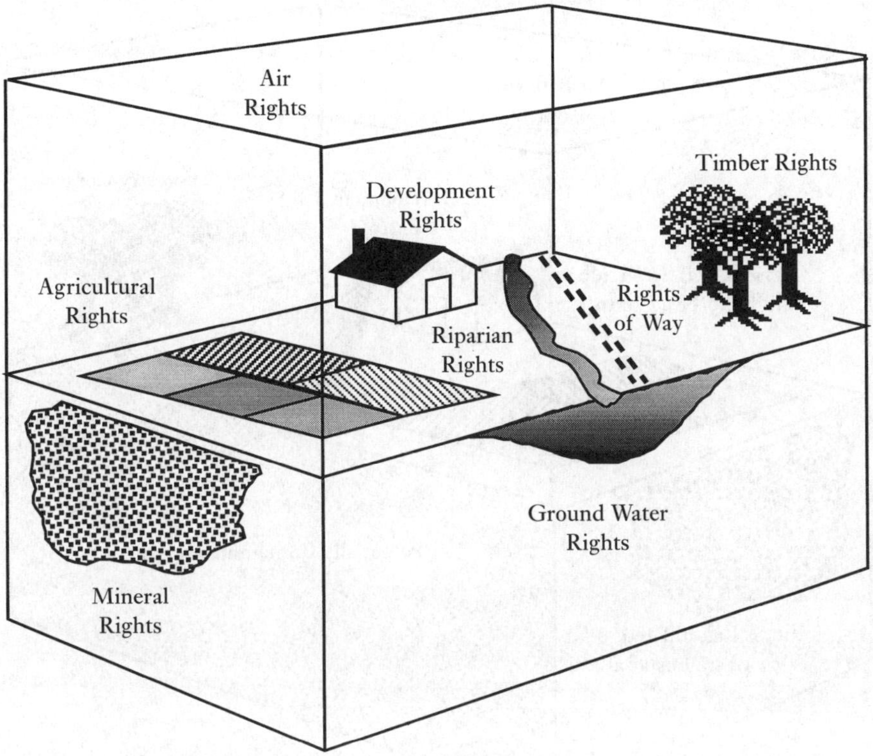

FIG. 1.3*a* The cadastral parcel and ownership rights

Source: Based on Platt 1975

through zoning mechanisms and the designation of areas of special interest, ranging from historic districts to fragile ecosystems.

The fiscal component focuses on the economic utility of the land. Its processes may be used to support increased revenue collection and production, and may act as incentives to consolidate or redistribute land or to use land for particular purposes.

Information management is integral to all three components described above: the juridical cadastre underpins land registration; the fiscal cadastre supports valuation and taxation; and zoning and other information systems facilitate planning and enforcement of regulations. Recognition that these components share common information requirements led to the concept of the multi-purpose cadastre as a community-oriented, parcel based system for integrating land related information collected and managed by different agencies (McLaughlin 1975). In recent decades there have been many efforts, for instance by Australian states and other jurisdictions, to develop information

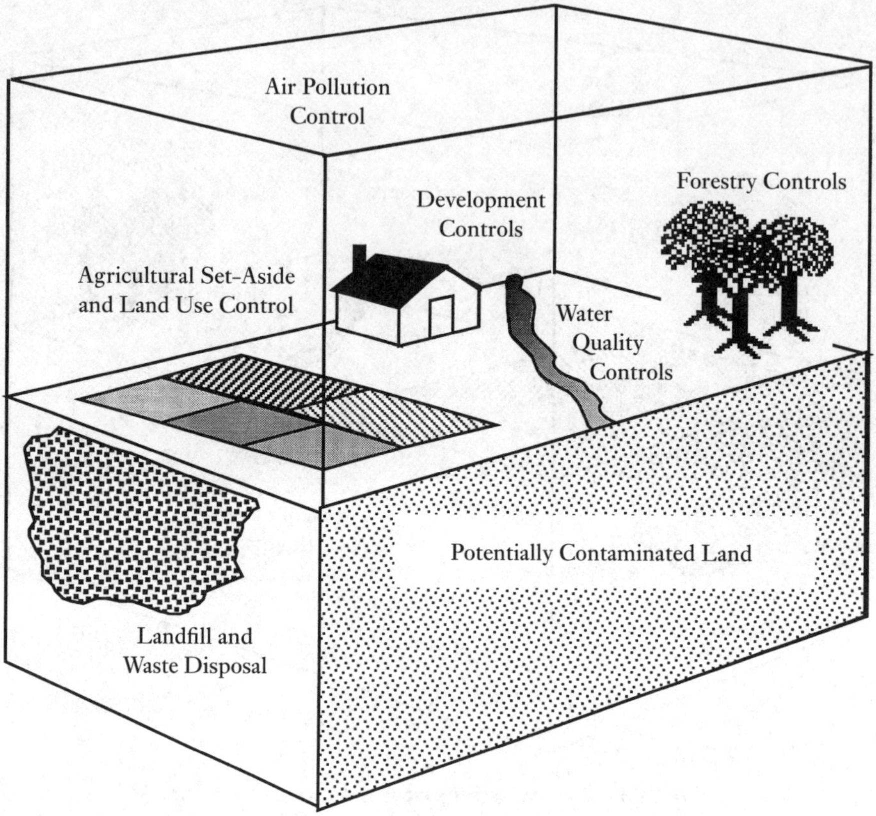

FIG. 1.3*b* The cadastral parcel and land use control (based on Platt 1975)

systems based on the cadastral parcel since this is the basic spatial unit of human activity.

The opportunities to develop land information systems have increased with recent advances in information technology, moves towards constructing national spatial data infrastructures (McLaughlin 1991; NRC 1993) and the creation of national land information services. While the function of a land administration system is to support the management of real property, including the physical earth and all things attached to it, the function of a land information system is to underpin this process.

1.8 Concluding Remarks

The present text builds on an earlier work by the authors on Land Information Management (Dale and McLaughlin 1988), seeking to broaden the debate

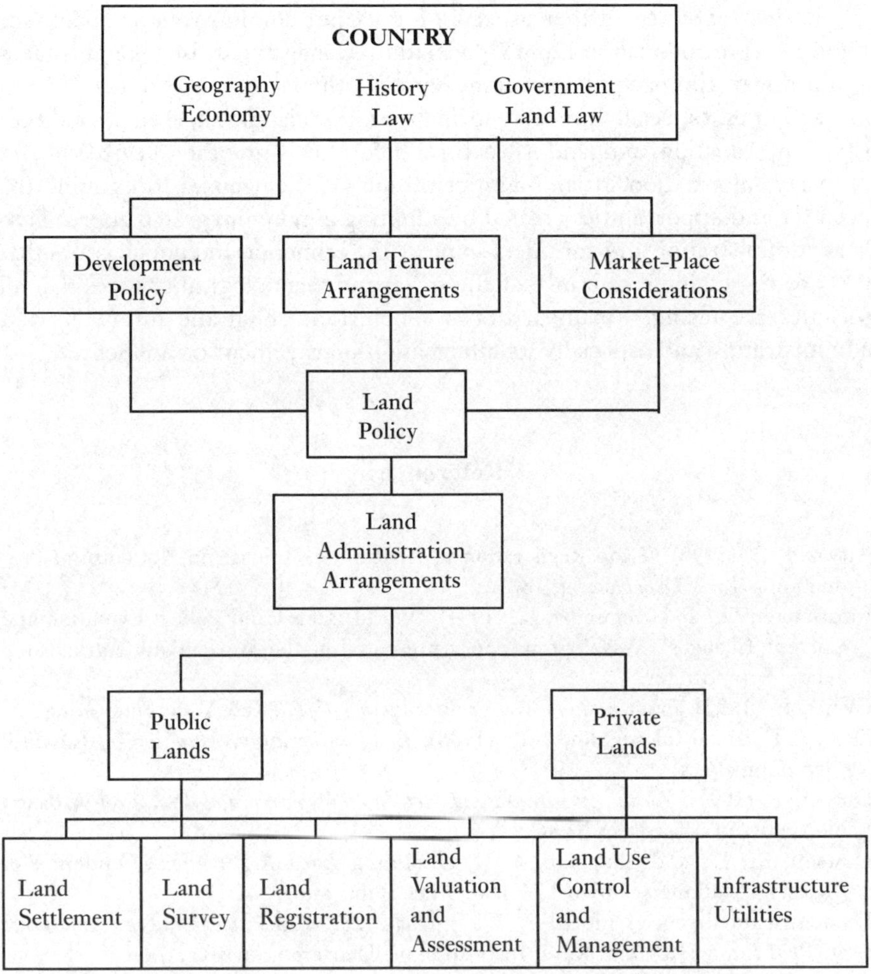

FIG. 1.4 Placing land administration in context

and focus more on policy than technical issues. In most cases deficiencies in land administration arrangements are due to institutional (including management) rather than technical problems. While the text focuses on the problems of under-resourced countries it also seeks to identify the bedrock on which all systems stand, both rich and poor. Each jurisdiction uniquely adapts land administration arrangements to accommodate its own history, culture, economics, and politics; nevertheless there are common elements and issues that all systems must address. The aim of this text is to identify both the differences and common matters that are important and of which decision-makers should be aware.

The text explores further aspects of the land administration model (see Figure 1.4) introduced in Land Information Management. In general it takes a 'top-down' approach by examining some of the factors that influence front line activities, especially property formalization, cadastral surveying, land registration, valuation, and land use control before revisiting the role of land and property information in land administration. It then moves to examine the need for and opportunities created by adopting a more integrated approach to land administration and considers some of the economic and policy issues that this creates, taking cognizance of the social and technological changes that all societies are facing. Finally it draws conclusions about the future of land administration and especially its information management component.

References

Attwood, D. (1990), 'Land Registration in Africa: The Impact on Agricultural Production', *World Development*, 18/5.

Binswanger, H., and Deininger, K. (1994), 'World Bank Land Policy: Evolution and Current Changes', Washington: paper presented at the World Bank Agricultural Sector Symposium.

Cymet, D. (1992), *From Ejido to Metropolis: Another Path* (New York: Peter Lang).

Dale, P. F., and McLaughlin, J. D. (1988), *Land Information Management* (Oxford: Clarendon Press).

Dorner, P. (1991), *Latin American Land Reforms in Theory and Practice* (Madison: University of Wisconsin Press).

Dowson, Sir E., and Sheppard, V. L. O. (1952), *Land Registration* (London: Her Majesty's Stationery Office, Colonial Office Publication No. 13).

Economist Intelligence Unit (EIU) (1997), *Global Direct Investment and the Importance of Real Estate* (London: Royal Institution of Chartered Surveyors).

Feder, G., Onchan, T., Chalamwong, Y., and Hongladarom, C. (1988), *Land Policies and Farm Productivity in Thailand* (Baltimore: Johns Hopkins University Press).

——and Feeny, D. (1991), 'Land Tenure and Property Rights: Theory and Implication for Development', *World Bank Economic Review*, 5/1.

Haugerud, A. (1989), 'Land Tenure and Agrarian Change in Kenya', *Africa*, 59/1.

Johnson, O. (1972), 'Economic Analysis: The Legal Framework and Land Tenure Systems', *Journal of Law and Economics*, 15/1.

McLaughlin, J. D. (1975), 'The Nature, Design and Function of Multipurpose Cadastres', Madison: Ph.D. Dissertation, University of Wisconsin.

——(1991), 'Towards National Spatial Data Infrastructure', *Proceedings of the 1991 Canadian Conference on GIS*, Ottawa, Canada.

——and Nichols, S. (1989), 'Resource Management: The Land Administration and Cadastral Systems Component', *Surveying and Mapping*, 49/2.

Migot-Adholla, S., Hazell, P., Blarel, B., and Place, F. (1991), 'Indigenous Land Rights Systems in Sub-Saharan Africa: A Constraint on Productivity', *World Bank Economic Review*, 5/1.

NRC (1993), *Toward a Coordinated Spatial Data Infrastructure for the Nation* (Washington: National Academy Press).

OECD (1986), *Labour Market Flexibility* (Paris: Organization for Economic Cooperation and Development).

Palmer, D. (1996), 'Incentive-Based Maintenance of Land Registration Systems', Gainesville: Ph.D. Dissertation, University of Florida.

Platt, R. (1975), 'Land Use Control: Interface of Law and Geography', Association of American Geographers, Resource paper, Washington, No. 75.1.

Putnam, R. (1992), *Making Democracy Work* (Princeton: Princeton University Press).

Shipton, P., and Goheen, M. (1992), 'Understanding African Landholding: Power, Wealth and Meaning', *Africa*, 62/3.

Simpson, S. R. (1976), *Land Law and Registration* (Cambridge: Cambridge University Press).

de Soto, H. (1993), 'The Missing Ingredient', *The Economist*, 11 Sept.

Swynnerton, R. J. M. (1955), *A Plan to Intensify the Development of Agriculture in Kenya* (Nairobi: Government Printer).

United Nations Centre for Human Settlement (UNCHS) (1996), *Habitat Agenda and Istanbul Declaration* (Nairobi: United Nations Centre for Human Settlement).

United Nations Economic Commission for Europe (UNECE) (1996), *Land Administration Guidelines: With Special Reference to Countries in Transition* (New York and Geneva: United Nations).

2

Land Tenure, Property, and Development

2.1 The Nature of Property and Rights in Land

The law provides a complex set of rules that have evolved within each society to ensure its orderly running and the peaceful behaviour of its members. The law may take several forms, amongst which broadly speaking, there is statutory law and customary law. Under statutory law, all rules and regulations are written down and codified; under customary law there is no written record but it is assumed that the code is well known by all members of society. In some jurisdictions there is the common law which grew out of customary law; over time the judgements of the courts have been written down and now create precedents whereby new cases can be judged. Many jurisdictions have legal regimes that combine in some fashion statutory and customary law.

The law of property deals both with relations between people (*in personam*) and of persons to things (*in rem*). The law recognizes different types of interest in property and makes a distinction between the physical objects and the abstract rights associated with their use. Land as real or immovable property is in many jurisdictions taken to include all things attached to it such as buildings and other permanent fixtures. It also usually includes the minerals below the soil and the air above, unless these are specifically excluded. In some countries, however, a distinction is made between 'land' as a natural object with soil and a surface and 'property', which is taken to mean the buildings and other man-made objects attached to the land. In the present context, land will be regarded as including all construction and development, while the word 'property' will normally be used more specifically to relate to the abstract nature of land.

Rights describe what may be done with property; they are abstract but none the less real in their effect. They have been described as being like a bundle of sticks associated with any property, one stick for each thing that can be done with the property. Rights provide the institutional foundation for addressing such questions as:

- Who owns the property either in whole or in part?
- In what ways may the property be used?

- Who may use the property and who may stop it being used?
- Who may examine the property and who may stop it being examined?
- Who may alter the property or otherwise add value to it?
- What legal liabilities exist for the quality of the property?
- Who is entitled to the profits from its use?
- What is a fair financial reward for investment in the property?
- Who can create new properties or prevent new properties from being created?
- Who can sell the property or give it away and who may stop its sale or gift?
- Who may buy the property?
- How may the property rights of owners and the public including third parties best be protected?

Ownership, the proof of which is known as 'title', is a fundamental legal concept. In common law jurisdictions, the highest form of ownership is freehold, which describes the complete bundle of rights that can be held privately at any point in time. On the death of a freehold owner, the rights pass to the heirs in accordance with the terms of the deceased's Last Will and Testament, or in the absence of such in accordance with laws governing intestacy.

Possession is both a legal concept and a matter of fact. In the case of a physical object, such as an automobile, possession may lie with a thief who has stolen it while ownership may lie with the person who purchased it and who has registered that ownership with the State. Indeed, at any given time, neither the thief nor the owner may have actual occupation of the automobile (it may, for example, be locked away in a garage).

In the case of real property, title does not necessarily imply occupation or use—the freehold owner of a house may have let the property in leasehold, retaining only certain residual rights. Within the duration of the lease, which might even be as long as 999 years, the freeholder cannot regain possession of the property unless there is a clear breach of contract. Similarly the freeholder may have pledged the property to a mortgage company and in default of payment may be forced to surrender the title.

2.2 Property in the Context of Land Tenure

Land tenure describes the manner in which rights in land are held. It is defined by a broad set of rules, some of which are formally defined through laws concerning property while others are determined by custom.

Throughout the world there are many different kinds of property regimes based on combinations of group, state, and individual property rights. Property regimes are often categorized as open access, communal or common property, private property, and state property (Feder and Feeny 1991). Open access occurs when there is no defined group of owners; hence the benefits are available to anyone and there are no duties or obligations. With common or communal property there is some aggregate body that has ownership of the whole and rights to exclude non-members; individual members of the body have both rights and duties with respect to the use and maintenance of the property. In the case of private property, individuals have rights of ownership, tempered by responsibilities and restrictions (placed by the state or by other third parties). With State property, public agencies set rules for access and use of the property and individuals have duties to respect these rules.

Most jurisdictions have some combination of these property regimes. Also, there is a hierarchy of property interests in land and these, together with the concomitant duties, obligations, and powers, comprise a dense network of intersecting interests known as the land tenure. In particular there are four broad categories of interest:

1. Overriding interests—such as the powers of expropriation held by the state.
2. Overlapping interests—for instance, the use of a parcel by more than one party through rights of ownership, easements, and leases.
3. Complementary interests—for example, rights held by several parties in the same property such as party walls and common areas in condominiums.
4. Competing interests—where different parties contest the same interests in the same parcel.

Tenure covers not only the rights to gift or grant land, buy and sell it, or mortgage it, but also the rights to use the land subject to certain restrictions or obligations. Land may belong to one person but be used by another. The most common form of this is renting through formal legal agreements. In some land tenure systems, land may be sold with certain rights being retained by the sellor (vendor); thus for instance the rights to fish in a stream that flows through a property may not be conveyed to the purchaser. The vendor may also in effect be able to eliminate some rights by refusing to pass them on.

In some jurisdictions it is possible to impose restrictive covenants on the land thus restricting the freedom of the purchaser—for example by insisting that a tract of land may only be used for residential purposes or may not be used for the sale of alcohol. Local municipality planning regulations and local by-laws may also impose such restrictions. Both public and private actions can therefore have a negative effect on how the land may be used.

In many cities, especially where they are expanding rapidly, access to land is a major problem. Squatter settlements spring up either as a result of forcible occupation or through the use of land for which there is no clear ownership.

In some jurisdictions, the peaceful occupation of land without formal legal agreement can lead to the prescription of rights by a process known in common law jurisdictions as adverse possession. The key to this is the peaceful nature of the occupation during which the use of the land goes unchallenged. If there is a challenge or if there are hidden agreements then adverse possession does not apply. Thus there may be 'peppercorn' rents where the occupier pays a trivial amount for the use of the land, a sum that is sufficient to acknowledge the rights of the formal owner. The doctrine of adverse possession allows land of unknown or uncertain ownership to be brought back onto the registers. The period of peaceful occupation necessary for prescription will be laid down in the law after which the occupier can make full claim to the ownership of the land.

For many people, especially the poor, the right to use the land may be more important than the right to ownership. In many of the land restitution programmes for countries in transition, a distinction has been drawn between ownership rights and use rights, the latter being easier to allocate and control by administrative decree. In such countries, private ownership rights may not have existed in recent times and the old registers and archives may need to be carefully scrutinized to determine the validity of any claims. Allocating use rights has allowed the occupation and development of the land to proceed without addressing either the legal concept of landownership or the political traditions that have claimed much or all of the land for the State.

Use rights may entitle the occupier to some or all of the profits that arise from using the property. In the case of land that is rented for commercial purposes the landlord would not normally anticipate a share of the profits from the business being conducted on that property, although to an extent the rent charged will reflect the expected profits that should be realized. In the Indian subcontinent, however, sharecropping has been common with the landlord taking anything from one-third to two-thirds of the profits as rent. Use rights often also form the basis for land taxation schemes (especially where it is too difficult or expensive to identify the actual owners).

Landowners frequently wish to restrict access to their land, for instance by recourse to the law of trespass that relates to illegal entry onto property. There is often a need to protect personal privacy and to respect the so-called territorial imperative that is recognized by many behavioural scientists. Restrictions may also be enforced for security reasons, either in the interests of the State (for example with regard to access to military compounds), Health and Safety (access to dangerous areas), or for the protection of removable property (bank vaults or shopping premises).

Land tenure systems determine the extent to which the rights in land and property may be transferred. Freehold owners may dispose of their land in whole or in part at will, subject to any restrictions that may exist. For example, in some countries, land may not be sold to foreign nationals; in such cases, foreigners may only be able to lease property or else must form partnerships with citizens. A leasehold owner may have rights to sell property for the outstanding duration of the lease or may only have limited rights including sub-letting.

2.3 Property in Support of Sustainable Development

Land tenure involves not only vendors and purchasers, owners and occupiers, but also third party interests, especially those concerned with the use of the land by future generations. This is a subject that emerged under the label of sustainable development as a major international policy issue with the publication in 1987 of the report of the World Commission on Environment and Development (WCED) entitled *Our Common Future*. The Commission had been organized on behalf of the General Assembly of the United Nations in order to 're-examine the critical environment and development problems on the planet and to formulate realistic proposals to solve them, and to ensure that human progress will be sustained through development without bankrupting the resources of future generations' (WCED 1987).

The term sustainability originally referred to harvesting regimes for specific stocks of natural resources that could be sustained over time (such as sustained-yield fishing). The use of the term has now evolved to refer to the challenges of (i) balancing the economic and ecological perspectives of development and (ii) acknowledging our collective responsibility of stewardship to future generations. According to the WCED:

> Sustainability requires the enforcement of wider responsibilities for the impacts of decisions. This requires changes in the legal and institutional frameworks that will enforce the common interest. The law alone cannot enforce the common interest. It principally needs community knowledge and support, which entails greater public participation in the decisions that affect the environment. This is best secured by decentralizing the management of resources upon which local communities depend, and giving these communities an effective say over the use of these resources. It will also require promoting citizens' initiatives, empowering people's organizations, and strengthening local democracy (WCED 1987).

In reviewing the role of property in supporting sustainable development, the World Commission report began by noting that ecological interactions do not respect the boundaries of individual ownership and political jurisdiction. An example cited was agricultural practices in watershed areas, where the ways in which a farmer up the slope uses land will directly affect run-off on farms downstream. However it went on to acknowledge that in many countries where land is very unequally distributed, land reform is a basic requirement. Without it, institutional and policy changes meant to protect the resource base can actually promote inequalities by shutting the poor off from resources and by favouring those with large farms, who are better able to obtain the limited credit and services available. By leaving hundreds of millions without options, such changes can have the opposite of their intended effect, ensuring the continued violation of ecological imperatives. (WCED 1987)

While much of the report focused on agricultural development, the Commission also addressed the urban challenge, which is that by the turn of the century more than half of the world's population will be living in urban areas. This will force government intervention to be reoriented so that limited resources can be concentrated on improving housing conditions for the poor. Seven priorities were identified:

1. Provide legal tenure to those living in 'illegal' settlements, with secure titles and basic services provided by public authorities.
2. Ensure that the land and other resources that people need to build or improve their housing are available.
3. Supply existing and new housing areas with infrastructure and services.
4. Set up neighbourhood offices to provide advice and technical assistance on how housing can be built better and cheaper, and on how health and hygiene can be improved.
5. Plan and guide the city's physical expansion to anticipate and encompass land needed for new housing, agricultural land, parks, and children's play areas.
6. Consider how public intervention could improve conditions for tenants and those living in cheap rooming or boarding houses.
7. Change housing finance systems to make cheap loans available to lower-income and community groups (WCED 1987).

Subsequent to the publication of the Commission's report, significant effort has gone into deepening people's understanding of the institutional mechanisms required to support sustainability, and the relationship between equity, good stewardship, and environmental resilience.

With respect to the role of property, there is an emerging consensus that property rights regimes 'to be effective in modulating the interaction between humans and their environment, must reflect both general principles and specific social and ecological contexts' (Hanna and Munasinghe 1995). Effective property rights regimes must specify both individual and collective rights and responsibilities; address the relationship between ecosystems and institutional structures that affect them; limit transaction costs; and create monitoring and enforcement processes at appropriate scales (Ostrom 1990; Bromley 1991).

2.4 Discrimination in the Access to Land

A key aspect of land tenure is the extent to which it inhibits access to land, a matter that has been a major focus of attention in recent years, especially within the United Nations agencies and international funding bodies. Discrimination exists against three broad groupings of people—women, indigenous peoples, and minority groups of settlers. In particular, the discrimination against women in many societies has been a matter of major concern, such discrimination often being built into the land tenure system. These concerns have in part to do with social justice and in part with economic development—many women in India for example cannot get access to credit to develop their cottage industries because their property rights are not formally recognized. There are even legal systems in which women are treated as minors, unable to enter into property transactions without a male relative's consent.

Furthermore, as noted in a recent United Nations Centre for Human Settlements publication, even in 'societies where women have the right to own and inherit land, the introduction of cash economies and the growing importance of land as a commodity results in women having little chance of owning land. This is due to the high cost of land, as well as women being unaware of their rights, where they exist' (UNCHS 1995).

In many developing countries, there may be no discrimination within the formal law; rather, restrictions on access to land and property arise through social custom and prejudice. In some countries, limited measures have been undertaken to improve the role of women in obtaining and managing property rights. In the Lao People's Democratic Republic land titling project, for example, provision has been made to train both management staff and adjudication teams in gender awareness. Staff have been encouraged 'firstly to ensure, so far as possible, that there is female representation on adjudication teams, with the current lack of balance being rectified when additional staff are recruited in the later years of the project. Secondly, the Lao Women's Union should be co-opted to assist in publicizing amongst women their rights in the land titling process' (BHP 1995).

In Kenya, the government has published a guide on land transactions for women, describing all the stages of land purchase and transfer. However, it still remains that 'fundamental change is needed to help women overcome barriers to land ownership. Land reform cannot be ignored in this process. To date, little research has been carried out on the question of women's access to land, and how the constraints they face could be addressed' (UNCHS 1995).

Restrictions on access to land and the rights to use land affect not only women. There is a major challenge in much of the world to reconcile the demands of aboriginal people for special status and rights with the existing institutional arrangements and ideological foundations of their governments (Dyck 1985). In many countries there are indigenous people whose formal rights have been overridden by settlers from abroad. In countries such as Australia with its Aboriginal people and New Zealand with its Maori, there have in recent times been high profile court cases seeking to restore the traditional rights of the native peoples. In Australia, for example, it was for a long time held that when the Europeans first arrived 200 years ago there were no people on the subcontinent (a doctrine known as 'terra nullius'). As far as the settlers were concerned, the Aboriginal people were regarded as unworthy of consideration. Only in the last decade have attitudes begun to change, partly as a result of what is known as the *Mabo* case in which the court held that the peoples living in the Torres Straits had rights in land that should and could be recognized. This has led to further reforms and more justice for the Aboriginal people.

On the Pacific islands of Fiji, there was a steady migration of peoples into the islands from the Indian subcontinent, resulting in the indigenous people becoming a minority in their parliament. The native Fijians then took matters into their own hands through a military coup. In other countries, the indigenous people have lacked the strength and cohesion to be able to assert their rights effectively. Hence the needs and concerns of peoples such as the Indians of North and South America, the Saami in Scandinavia, and the Dyaks in Borneo have to date received only marginal consideration. Migrant populations, such as the gypsy people in Europe, have suffered equally badly. The problems tend to become acute when areas of traditional occupation become targets for development or natural resource exploitation, for instance for mineral or timber extraction.

The Rio Declaration on the Environment and Development (UN 1992) adopted a principle (Principle 22) dealing specifically with the indigenous issue: 'Indigenous people and their communities and other local communities have a vital role in environmental management and development because of their knowledge and traditional practices. States should recognize and duly support their identity, culture and interests and enable their effective participation in the achievement of sustainable development.'

To achieve the goals of sustainability and the constructive participation of indigenous people, there are needs among many other things for:

- Recognition—of the special values, culture, practices, and knowledge of indigenous peoples.
- Integration—of traditional knowledge and practices of indigenous peoples into contemporary management systems.
- Involvement—of indigenous peoples in decisions about their environment and development.

- Targeting—of specific initiatives relevant to indigenous peoples, such as dealing with land tenure issues.
- Capacity building—amongst governments dealing with indigenous issues, such as through the training of government staff (Cicin-Sain and Knecht 1995).

Canada provides an example of a jurisdiction wrestling with the issues of indigenous property rights. Currently new models for aboriginal governance and new property regimes are being implemented over more than a million square kilometres in northern Canada. The Canadian constitution recognizes the inherent right of the aboriginal peoples of Canada to govern themselves 'in relation to matters that are internal to their communities, integral to their unique cultures, identities, traditions, languages and institutions, and with respect to their special relationship to their land and their resources' (Government of Canada 1995).

Discrimination over access to land is not however confined to indigenous peoples; the reverse is often the case, most noticeably in much of east and central Europe. In many of the former communist countries in Europe there are ethnic groups whose presence is now not wanted. In Poland, for example, there are fears that former German settlers who had previously and at the time quite legitimately built their homes in areas then known as Prussia will wish to reclaim their lands as part of the restitution programme; in the Sudetenland of the Czech Republic there are similar fears. In applying for membership of the European Union, such countries must show that they will allow all citizens of the Union equal rights of access to their land markets. Under the Maastricht Treaty, Denmark negotiated certain exceptions to this and was allowed to restrict the sale of summer cottages on the coastline of Jutland where non-Danes were increasingly buying up the land. The new applicants for membership of the European Union are hoping for similar special treatment.

In Latvia, 70 per cent of the inhabitants of the capital city Riga are of Russian origin, with a figure of 30 per cent for the country as a whole. One-third of the indigenous population were deported to Siberia after the Second World War and were replaced by Russian settlers. Having recently regained their independence, such countries are reluctant to surrender their sovereignty to non-ethnic citizens who may at present have much greater economic power or be capable of greater influence than are the people of local ethnic origin. Many people are conscious of what has happened in parts of the former Yugoslav Republic where in recent times there have been bloody battles between ethnic groups over land.

There are no easy solutions to the problems of civil strife, especially when they relate to access to land. Establishing the facts and awareness of the sensitivity of the issues must however be parts of the way forward.

References

BHP (1995), *Lao People's Democratic Republic Land Titling Project: Project Preparation Report*, Washington: Report prepared for the World Bank, Report No. DOLHM 1/1995.

Bromley, D. W. (1991), *Environment and Economy: Property Rights and Public Policy* (Oxford: Basil Blackwell).

Canada, Government of (1995), *Aboriginal Self-Government* (Ottawa: Federal Policy Guide).

Cicin-Sain, B., and Knecht, R. (1995), 'Analysis of Earth Summit Prescriptions on Incorporating Traditional Knowledge in Natural Resource Management', in *Property Rights and the Environment: Social and Ecological Issues* (Washington: World Bank and the Beijer International Institute of Ecological Economics).

Dyck, N. (ed.) (1985), *Indigenous Peoples and the Nation-State* (St John's: Memorial University of Newfoundland publication).

Feder, G., and Feeny, D. (1991), 'Land Tenure and Property Rights: Theory and Implication for Development', *World Bank Economic Review*, 5/1.

Hanna, S., and Munasinghe, M. (eds.) (1995), *Property Rights and the Environment: Social and Ecological Issues* (Washington: World Bank and the Beijer International Institute of Ecological Economics).

Ostrom, L. (1990), *Governing the Commons* (Cambridge: Cambridge University Press).

United Nations (UN) (1992), *Agenda 21: Rio Declaration*, Rio de Janeiro: Final text of agreements negotiated by governments at the United Nations Conference on Environment and Development.

United Nations Centre for Human Settlements (Habitat) (1995), *Women in Human Settlements Development* (New York: UNCHS).

World Commission on Environment and Development (WCED) (1987), *Our Common Future* (Oxford: Oxford University Press).

3

Formalizing Property Rights

3.1 Formal and Informal Property

Property systems may be formal or informal. Formal property systems are those where the interests are explicitly acknowledged and protected by the law. This is the case for the vast majority of property rights in developed countries. Informal property interests are those that are recognized by the local, informal community but which are not formally acknowledged by the state. They exist in most developing countries outside the legal system and are often the result of inadequate legislation, or excessive and inefficient bureaucracies.

Many legal systems, such as those based on the French Napoleonic code, have been established 'top down' with a framework of law imposed by legislators. The common law systems on the other hand are based on a 'bottom-up' approach in which the customs and practices of the people eventually become written down and accepted within a statutory framework. Historically, common law systems grew out of informal systems and, through the body of case law that developed, gradually became accepted across the whole of the jurisdiction. 'Top-down' legal systems are essentially negative in that actions may not be undertaken unless they are permitted by the law; 'bottom-up' systems generally work on the basis that anything is permitted unless explicitly forbidden by the law. In many of the central and eastern European countries, land reform has been delayed because there were decisions that could not be taken because there was no law that permitted them. Rather than move on with the processes, laws had to be drafted and agreed specifying that such actions were permissible. As an example, work could not be contracted out to the private sector because the law did not say that this was permissible; there was however no statement that such action was forbidden.

Informal systems of tenure provide no state security but can, in practice, be sufficiently robust for the people in the areas concerned to invest in housing and development; an estimated three-quarters of Greater Cairo, for example, is said to have been developed without formal approvals. In general, formal systems give rise to security, provided that they are properly administered; where the formal system is poorly managed or where it does not have the support of the majority of the people because it is seen as a mechanism to protect the elite, security may break down.

Insecurity differs widely in different property regimes (Wachter 1996). Problems of land tenure insecurity differ substantially for example between areas where there are private property rights (as in a Latin American agrarian structure), where there is common property (such as in a traditional African land system) or state property (for instance in the state farms in former socialist countries), or in open areas (such as squatting in tropical forests).

3.2 The Case for Property Formalization

In Latin America, the concept of property formalization has emerged as a key strategy for dealing with tenure insecurity. As countries move towards market-oriented democracies, most Latin American governments have recognized the goal of widespread property ownership. However, they have often failed in practice to recognize many property rights already held by their citizens. In Peru, for example, about 90 per cent of rural parcels and 60 per cent of those in urban areas are informally owned (de Soto 1993). Similarly, in the Brazilian city of Recife, it is estimated that more than 60 per cent of the population live in squatter settlements occupying no more than 10 per cent of the city's area (Maia 1995). Property may be informal because it is illegal, by which is meant that it is held in direct violation of the law. This may be as blatant as occupation of a site in contravention of an eviction notice. Frequently, however, rights may be illegal simply because the laws are inappropriate.

Property may also be informal because it is extra-legal. In such cases property rights may be held with the tacit consent of the state even though formal acknowledgement is withheld. New city dwellers may live without being harassed by the state, but their property may still lack formal recognition for reasons ranging from political paternalism to bureaucratic inertia. As a result, many of those in the informal sector have become 'second class' citizens. Peru offers an example of such a system where migration to urban areas increased so dramatically that successive governments conceded defeat and stopped removing squatter settlements on vacant state land surrounding Lima.

Regardless of their origins, informal rights do not exist merely because someone claims them. Instead, people within the informal community agree among themselves as to where and how each can exercise these rights. The social basis of informal ownership is often clearly defined. In large measure, people are encouraged to operate in the informal sector when the costs of legal transactions increase as a result of the failure of institutions to facilitate exchange and enforce contracts (de Soto 1989).

Frequently, attempts to formalize property have been unsuccessful because they have ignored local property rules and replaced them with foreign ones. In particular, titling and registration projects in developing countries have in the past almost invariably been associated with individualization of private property and have used processes and mechanisms imported from outside. This has arisen in part from a form of institutional imperialism: western consultants and aid organizations tend to export systems with which they are familiar.

Imposing similar rules in societies with different institutional arrangements has led to widely divergent outcomes since 'although the rules are the same, the enforcement mechanisms, the way the enforcement occurs, the norms of behaviour, and the subjective models of the actors are not' (North 1990).

Simpson (1976) believed that formalizing property rights should not change the customary property system; in his view, adjudication recognizes existing rights but does not create new ones. He saw registration merely as a device to give certainty of ownership of land and to enable dealings to be conducted quickly and cheaply. Although often cast as a proponent of the 'westernization' of indigenous property rights, Simpson offered provisions for registering family (or group) rights and for designating areas of 'customary land'.

In practice, adjudication and title registration are rarely benign and often alter the status quo. As Shipton and Goheen (1992) have noted, informal property systems can seldom be legislated away in favour of formal ones; hence creating new legal structures adds layers of procedures on to others that are unlikely to disappear. The resulting tension between formal and informal rights to land may actually reduce overall security and increase the number of disputes; if left alone, informal property systems may on their own provide considerable security of tenure (Haugerud 1989; Migot-Adholla et al. 1991; Binswanger and Deininger 1994). Furthermore, credit may not flow as easily as expected to newly formalized property owners since the lack of a robust banking system will restrict the ability of commercial banks to make loans. Experience also suggests that credit from state-controlled banks is often diverted to those with political influence.

Some writers such as Attwood (1990) have concluded that the human, technical, and financial resources required to establish and maintain a formal system of land administration in a developing country may be better applied to other important sectors or activities. However, increasing evidence from countries such as Peru and Thailand suggests that benefits can accrue when laws are made to fit everyday life and not the other way around, and when costs of formal transactions are reduced (Feder and Nishio 1998).

3.3 Land Reform and Formalization

The term 'land reform' covers a variety of processes of which property formalization may be one; others include legislative reform, land tax reform, and land market reform. Three basic types of land reform programme may be identified:

1. those mainly of a legal, technical and/or administrative nature, carried out primarily to make the infrastructure work more effectively (for instance land consolidation, legislative or tax reform);
2. those designed to deal with the more basic concerns for minimizing tenure insecurity (such as property formalization programmes);
3. those designed to tackle the even more fundamental challenge of inequality of ownership (such as land redistribution or settlement).

The starting point for all land reform programmes is the determination of existing formal and informal rights in land. In many of the former European communist countries, land restitution has been concerned with the formal rights that officially existed prior to the time when the communists took over. Elsewhere land reform has begun either with the formal rights that operated under a less efficient land administration system (as in the reforms in the 1970s and 1980s in many of the Caribbean islands and, for instance, in Thailand) or with the previously unrecorded rights that were held under traditional forms of tenure (as in many African countries).

Land formalization will normally begin with a process known as adjudication of title to land in which existing rights in parcels of land are authoritatively determined. This necessitates determining 'who' owns 'what', that is the rights and ownership must be ascertained as well as the extent of the land affected. The latter means that the boundaries of each parcel must be agreed between the adjoining parties. In particular, adjudication is the first stage in the registration of title to land in areas where the ownership of the land has not officially been determined. It is also a prerequisite for land consolidation and redistribution in order to ensure that the owners of each parcel of land are identified and treated fairly.

The land adjudication process may operate sporadically, by which is meant 'here and there' or 'now and then', or systematically. Sporadic adjudication takes place in a piecemeal, haphazard, and unpredictable sequence whenever or wherever there is a need to determine the precise ownership and limits of individual parcels—for example when a transfer of rights is about to take place or when an owner so requests. Systematic adjudication takes place in a logical and orderly fashion, for instance street by street or village by village, so that the rights to all land within a pre-defined area are determined at the same time.

Sporadic adjudication can be used selectively to encourage specific categories of landowners, for example progressive farmers, to take advantage of the system of registration of title. In the short term, sporadic adjudication is usually cheaper than systematic because decisions on the rights to many land parcels can be deferred. Hence less capital investment is necessary. The approach also permits the cost of the whole operation to be passed directly to the beneficiaries who can be charged an appropriate fee for having their land registered.

Systematic adjudication on the other hand is, in the longer term, less expensive because of economies of scale. It is more secure because it gives maximum publicity to the determination of who owns what within an area, and more certain because detailed investigations take place on the ground on the basis of direct evidence from the owners of adjoining properties. It has the advantages of publicity, on-the-ground inspection, easy administration, the possibility of using less expensive survey methods and the opportunity to link the process with land reforms, including land consolidation.

Whether undertaken through a sporadic or systematic approach, adjudication of title to land is a legal process. It must take place in accordance with specific legislation that identifies the procedures to be followed. The legislation should also specify appeals procedures such that, within a defined period, the judgements made at the time of adjudication can be either queried or else finalized; if after sixty days, for example, there is no appeal then the decisions can be entered finally in the registers.

For systematic adjudication to be successful there must be wide publicity in advance of the decision-making such as to ensure that landowners and occupiers understand fully the significance of what will take place and when it all will happen. Administrators need to decide which areas will be treated as a priority—for example areas that are to be subject to land reform, are under rapid development, have a high level of ownership or boundary disputes, or where there is need for credit.

Where a register of deeds is already in existence and the intention is to convert it into a register of titles, recourse to adjudication in the field may be avoided. If there is adequate mapping of the physical boundaries, then careful examination of the deeds and supporting documents may be sufficient to identify the parcels and their associated sets of property rights. If, however, the physical evidence is at variance with the documentary, then investigation on the ground may still be necessary.

Searching through old archives to find documentary evidence of previous ownership can be very time-consuming. When several of the eastern European countries began their land restitution programmes it took over a year to check a very small number of titles. Although over time the processes have been speeded up, they have not been fast enough to satisfy the demands of the aspirant landowners. As a result, many countries have been forced to issue land use certificates to ensure continuing agricultural production. Since in many cases there has been no opportunity to use the land as collateral, partly because there was no legislation to cover the mortgaging of land and partly because few banks were in a position to provide the necessary finance, this has proved an acceptable interim process.

A different historical context resulted in a very different approach in Peru where, since the 1950s, there has been dramatic urban growth, much of which is attributed to migration from rural areas. As people flocked to the cities, especially Lima, informal settlements developed at a rapid pace; such settlements scarcely existed in the 1940s. By 1972, one-quarter of the population in Lima occupied land informally and by 1985 more than two-thirds of housing stock was informal. Although Peruvian property legislation was progressive compared to that of many other developing countries, the formalization process was costly, complex, and time-consuming. An applicant could wait twenty years before receiving title from the government. Furthermore, titles were often not recorded in the land registry ('Registro de la Propiedad Inmueble') because of its burdensome registration requirements; furthermore, titles granted by the state were frequently flawed and lacked the legitimacy needed for use as collateral and related purposes.

Many of the informal settlements occurred on barren state land surrounding the old city of Lima. Some took place as the result of a carefully prepared invasion of underutilized land, planned well in advance by the leaders of the invasion. The site to be invaded was selected according to criteria such as topography, access, and the ease with which the land could be secured. Leaders recruited families to participate in the settlement since a large population at an early stage reduced the risk of eviction. Plans showing streets and parcels for residences, open spaces, and public buildings were drawn—often by engineering students—and members of the group were assigned their parcels. All this was done with great secrecy so that the authorities could not prepare counter-measures.

The date for the invasion was often a civil or religious holiday, thus giving the settlers time to consolidate their position. On the night of the invasion, street and parcel boundaries were marked on the ground, and families began constructing their shelters using straw matting. As soon as possible, residents built brick walls along the parcel boundaries to avoid disputes. Over the years, residents would replace their straw houses with ones of brick as their level of security increased and in accordance with their financial and labour resources and priorities. People bought and sold these parcels using informal titles.

Democratically elected local leaders administered these settlements long before Lima's formal local governments were similarly elected. Each informal settlement's local government was responsible for internal matters, including land administration (employing rudimentary registers), and initiating community improvement projects (such as building schools, community centres, and roads). In addition, it represented the settlement in any dealings with the police and in attempts to win formal recognition for the settlement from government agencies.

Beginning in 1961, the Peruvian government passed a series of laws aimed at legalizing the existing settlements and allowing people to gain formal recognition of their informal rights. An informal settlement attempting to become formal follows the reverse order of processes used to develop a formal settlement. In existing formal areas, title to the land is legally acquired, houses are then built, services such as roads, water, and sewage disposals systems are provided, and the residents then move in.

In informal settlements, residents occupy the land first, build their houses, slowly acquire services, and finally receive their titles. There is, however, a complex relationship between building a house and acquiring interests in the land. The greater the feeling of security enjoyed by a family, the more inclined it is to invest in home improvements, while at the same time having a robust house reduces the risk of eviction since it is harder to destroy.

In the mid-1980s Hernando de Soto and his 'Instituto Libertad y Democracia' (ILD) identified informal property institutions and legal and bureaucratic obstacles preventing formalization of rights recognized by informal communities. Based on this analysis, the ILD drafted a comprehensive reform of 175 laws and 2,000 regulations that in one way or another affected the property formalization process. Moreover, the ILD designed an efficient registry system ('Registro Predial') for registering newly formalized property rights and subsequent transactions. A set of reform laws and regulations drafted by the ILD became effective between 1988–90. The ILD tested the reform by formalizing 150,000 parcels between 1990–3.

The reform has reduced the costs of making informal parcels formal. It has widened the door to formal society by recognizing certain proofs of ownership used in informal communities. It has made the process more efficient by enabling all property holders in a community to formalize their properties at the same time by using common maps and simple forms. It has also allowed ownership and boundary information to be managed efficiently using a low-cost, parcel based computer registry system.

The reform has also reduced a bottleneck that was caused by having a small number of registry officials examine documents, by enabling private sector lawyers and engineers to take responsibility for verifying a document's legality before it reaches the registry office.

3.4 Land Titling

As decolonization progressed through the 1960s and governments became more conscious of the benefits of secure title to land, the need for improved land titling moved higher up the political agenda. By the early 1980s, thanks largely to the work of people like Harold Dunkerley, the World Bank organized a series of high-level international workshops that addressed land titling issues. Many projects were initiated, the typical components of which included:

- improvements to the legal framework and statutes;
- regularization of possessory titles (adjudication and registration);
- cadastral surveying and mapping;
- information technology support;
- institutional strengthening and project management support; and
- human resources development.

International financial institutions such as the World Bank, the Asian Development Bank, and, since the beginning of the 1990s, the European Union, have tended to focus on rural development projects. Programmes in Brazil and Central America and in east and central Europe have been primarily concerned with improving agriculture and the standards of living of the rural communities. Urban development projects have often ignored land tenure reform and many have suffered in consequence. One major example of an urban land titling project currently underway is the Peruvian Urban Formalization Project being funded by the World Bank (McLaughlin and de Soto 1994).

A few projects have bridged the gap between urban and rural land resource management—for instance in Thailand where the World Bank has for twenty years been a partner in a project to issue registered titles to every eligible rural parcel holder in the country and to reform landholdings in the capital city of Bangkok. (The Thailand Land Titling Project was a co-winner of the first World Bank Award for Project Excellence in 1997 (Rattanabirabongse et al. 1998).) Another example of a jurisdiction-wide initiative is the Greek Cadastre Project, a huge, ambitious undertaking designed to ultimately register 20 million urban and rural properties. Nevertheless, most projects, especially those funded by the European Union in east and central Europe, have focused on rural productivity and agricultural support, on linkages with rural credit schemes and on regional economic growth strategies.

3.5 The Benefits of Land Formalization and Land Titling

The benefits of land formalization and the introduction of a land titling system underpinned by a sound cadastral survey have been recited elsewhere (Dale and McLaughlin 1988). They include:

- Certainty of ownership not only as to who are the landowners but also what other rights exist in the land. This in turn should lead to greater social cohesion.
- Security of tenure since through the adjudication process, existing defects in any titles to land can be cured by the judicious use of appropriate powers. In many countries the official record is supported by a State guarantee of the title to the land so that anyone adversely affected by errors in the register may receive compensation. Greater security should in turn lead to greater productivity from the land, especially in rural areas where farmers have an incentive to manage their land more carefully and to invest capital and resources in it.
- Reduction in disputes over land and boundaries. Such disputes can give rise to expensive litigation, clog up the court system, and are socially divisive.
- Improved conveyancing so that the expense and delays in transferring property rights can be substantially reduced. Duplication of effort, for instance in the repeated investigation of old titles, can be avoided thus saving on costs.
- Security for credit since land and property are often used as security against any loan. Evidence suggests that the combination of a sound title with the ability to raise long-term credit can give rise to substantial increase in productivity from the land.
- Stimulation of the land market by not only lowering transaction costs but also by permitting the market to respond effectively to all the needs of users.
- Monitoring of the land market, leading to improvements in valuation and assessment and hence in property portfolio management.
- Facilitating of land reform such as land redistribution, land consolidation, and land assembly for development and redevelopment.
- Management of state land so that the state as a major landowner can husband its own resources and ensure that land leased from the state is properly controlled.
- Support for land taxation and improving the rate of revenue collection through a clearer identification of all land parcels and their owners.
- Improvements in physical planning in both the urban and rural sectors ensuring that development programmes can proceed in the certain knowledge of existing land rights.
- Recording of land resource information through the development of the multi-purpose cadastre (Williamson 1986).

However, such benefits only arise when formalization and land titling are seen as more than the technical and legal processes of facilitating land transfer. They have a long-term impact on economic and social development that goes well beyond simply improving the land conveyancing procedures.

References

Attwood, D. (1990), 'Land Registration in Africa: The Impact on Agricultural Production', *World Development*, 18/5.

Binswanger, H., and Deininger, K. (1994), 'World Bank Land Policy: Evolution and Current Changes', Washington: Paper presented at the World Bank Agricultural Sector Symposium.

Dale, P. F., and McLaughlin, J. D. (1988), *Land Information Management* (Oxford: Clarendon Press).

Feder, G., and Nishio, A. (1998), 'The Benefits of Land Registration and Titling: Economic and Social Perspectives', *Land Use Policy*, 15/1.

Haugerud, A. (1989), 'Land Tenure and Agrarian Change in Kenya', *Africa*, 59/1.

McLaughlin, J., and de Soto, H. (1994), 'Property Formalization: The PROFORM Solution', *Geomatica*, 48/4.

Maia, M. L. (1995), 'Land Use Regulations and Rights to the City: Squatter Settlements in Recife, Brazil', *Land Use Policy*, 12/2.

Migot-Adholla, S., Hazell, P., Blarel, B., and Place, F. (1991), 'Indigenous Land Rights Systems in Sub-Saharan Africa: A Constraint on Productivity', *World Bank Economic Review*, 5/1.

North, D. (1990), *Institutions, Institutional Change, and Economic Performance* (New York: Cambridge University Press).

Rattanabirabongse, V., Eddington, R., Burns, A., and Nettle, K. (1998), 'The Thailand Land Titling Project—Thirteen Years of Experience', *Land Use Policy*, 15/1.

Shipton, P., and Goheen, M. (1992), 'Understanding African Landholding: Power, Wealth and Meaning', *Africa*, 62/3.

Simpson, S. R. (1976), *Land Law and Registration* (Cambridge: Cambridge University Press).

de Soto, H. (1989), *The Other Path* (New York: Harper & Row).

—— (1993), 'The Missing Ingredient', *The Economist*, 11 Sept.

Wachter, D. (1996), *Land Tenure and Sustainable Management of Agricultural Soils* (Berne: Centre for Development and Environment).

Williamson, I. P. (1986), 'Cadastral and Land Information Systems in Developing Countries', *Australian Surveyor*, 33/1.

4

Land Registration

4.1 The Function of Land Registries

Land registration systems provide the means for recognizing formalized property rights, and for regulating the character and transfer of these rights. Registries document certain interests in the land, including information about the nature and spatial extent of these interests and the names of the individuals to whom these interests relate. They also normally record charges and liens, that is rights to retain property against debts as in the case of mortgages, although in some systems these are held in separate registries. In addition, land registries provide documentary evidence that is necessary for resolving property disputes as well as information for a wide variety of public functions (such as land valuation).

There are at least three basic types of land registration system: (i) private conveyancing; (ii) registration of deeds; and (iii) registration of title. Under a private conveyancing system, land transactions are handled by private arrangement. Interests in land are transferred by the signing, sealing, and delivery of documents between private individuals with no direct public notice, record, or supervision. The pertinent documents are held either by the individuals to a transaction or by an intermediary such as a notary. In such a system, the state has little control over the registration process (save for regulating the intermediaries) and there is little if any security for errors or fraud. Also, private conveyancing systems are invariably slow and expensive. Despite these serious limitations, notarial versions of private conveyancing are still found in many parts of Latin America.

4.2 Registration of Deeds

Under a system of registration of deeds, a public repository is provided for registering documents associated with property transactions (deeds, mortgages, plans of survey, etc.). There are three basic elements in deeds registration: the logging of the time of entry of a property document; the indexing of the instrument; and the archiving of the document or a copy thereof. While there are many types of deed registration system, they are all based on three core principles (Nichols 1993):

1. Security—registration of a document in a public office provides some measure of security against loss, destruction, or fraud.
2. Evidence—registered documents can be used as evidence in support of a claim to a property interest (although they cannot provide an assurance of title).
3. Notice and priority—registration of a document gives public notice that a property transaction has occurred and, with exceptions, the time of registration provides a priority claim.

The limitations of the system are well known and have been documented by such authors as Simpson (1976). Deeds registration provides a means for registering legal documents only; it does not register title to a property. Registration is often not compulsory and, as a general rule, many rights are not registered. Reviewing and assessing all the documents required to determine the validity of a claim to ownership (the so-called chain of title) can be extremely tedious and expensive to undertake and is always open to dispute.

As noted in Dale and McLaughlin (1988), there are many ways of improving the deeds registration process, such as:

1. better basic records management, including better administrative and accounting arrangements;
2. standardization of forms and procedures to expedite the routine processing of documents;
3. physical improvements to the record keeping and the manner in which documents are stored so that they can be more easily accessed;
4. the use of microfilm both for archiving and retrieval of data;
5. more realistic and more flexible standards for surveying and mapping so that cheaper survey methods become possible;
6. partial examination of cadastral surveys and greater flexibility in the methods used;
7. partial examination of title documents, sampling rather than fully scrutinizing their contents, and hence taking a reasonable degree of risk;
8. compulsory registration so that there is a complete record of all transactions in land;
9. automation of the indexes to provide quicker document retrieval; and
10. computerization of the Abstracts of Title to provide quicker access to the important pieces of information.

4.3 Registration of Title

Registration of title was designed to overcome the defects of deeds registration and to simplify the process of executing property transactions. In such a system the register describes the current property ownership and the outstanding charges and liens. Registration is normally compulsory and the state plays an active role in examining and warranting transactions.

In most countries the entry on the register becomes the proof of ownership. According to Burdon:

> Registration of Title seeks to make a definitive statement as to the nature and extent of title, the land being identified by reference to a map. In registration of title an official examines the progress of deeds relating to a property and makes up a formal certificate of title beyond which no further examination need, in principle, be made. The extent of the property is plotted onto a map, and the property is allocated a unique identifying number. Except for any 'over-riding interest' as defined in statute, the formal title sheet is determinative of title. In most cases, title, once issued, is indefeasible. Should someone with a better claim to the land emerge and establish their claim they do not recover 'their' property but rather have a remedy against the State in indemnity. (Burdon 1998)

There are various types of title registration system, the best known of which is that introduced by Sir Robert Torrens in Australia in the latter half of the nineteenth century. The Torrens registration system is based on three well-known principles:

1. The mirror principle—the register reflects accurately and completely the current state of title; hence there is no need to look elsewhere for proof of title.
2. The curtain principle—the register is the sole source of title information. In effect a curtain is drawn blocking out all former transactions; there is no need to go beyond the current record to review historical documentation.
3. The insurance principle—the state is responsible for the veracity of the register and for providing compensation in the case of errors or omissions, thus providing financial security for the owners.

Title registration systems are in general supported by comprehensive surveying and mapping programmes that provide for parcel indexes and precisely delimited property boundaries.

Title registration represented a significant improvement over the rudimentary deeds registration systems of the nineteenth century. The process has been criticized for being expensive and cumbersome to implement and for the length of time required for state examination and approval of the title and survey evidence. In particular, delays often arise because of emphasis that is given to precise boundary delimitation. Furthermore the introduction of overriding interests that have a legal effect on the property although they are not specifically registered gradually diminishes the significance of the mirror principle as the details on the title certificate do not reflect all the rights as they exist on the ground.

4.4 Recent Analysis of Land Registration

Land registries are often perceived as complex and archaic operations that are captive to the property professionals who work with them. In the view of many lay people, the regulations seem to have been set in place and are maintained not to protect the public but to protect the privileged positions of these professionals (Stigler 1971). Also, land registration systems have traditionally had little influence on land or economic policy and have provided little opportunity for serious revenue collection. This is beginning to change however as jurisdictions look towards restructuring their registration systems, integrating these with other property-related activities, and running the registries in a more businesslike fashion.

Nichols (1993) has provided a comprehensive examination for upgrading and strengthening the registration function—see Figure 4.1. Beyond this, the traditional ways of differentiating between registry systems reviewed above are increasingly viewed as limiting and tending to distort the ways of classifying the registry function. Initial efforts to build new frameworks included McLaughlin (1975) and Dale and McLaughlin (1988); both employed an information management paradigm.

More recently Palmer (1996) has advanced a different way of thinking about the registry function based upon five criteria: jurisdiction-wide coverage, quality control, currency, guarantee, and indemnification.

The first of these, jurisdiction-wide coverage, suggests that the more parcels that are registered in a formal registry, the more effective will be the system. A 'seamless' property fabric covering an entire jurisdiction enables better resource management decisions to be made. From an economic perspective, small markets do not generate as much growth as those in which there are many participants while, from a social justice perspective, benefits of formal property should be provided to the majority in the country. The traditional approach to achieving jurisdiction-wide coverage has been through compulsory registration with no transaction being valid unless it is duly registered.

Compulsory registration, however, cannot work successfully without some means of enforcement and this is often lacking in low-income countries. Enacting a rule that registration is compulsory will often merely promote a

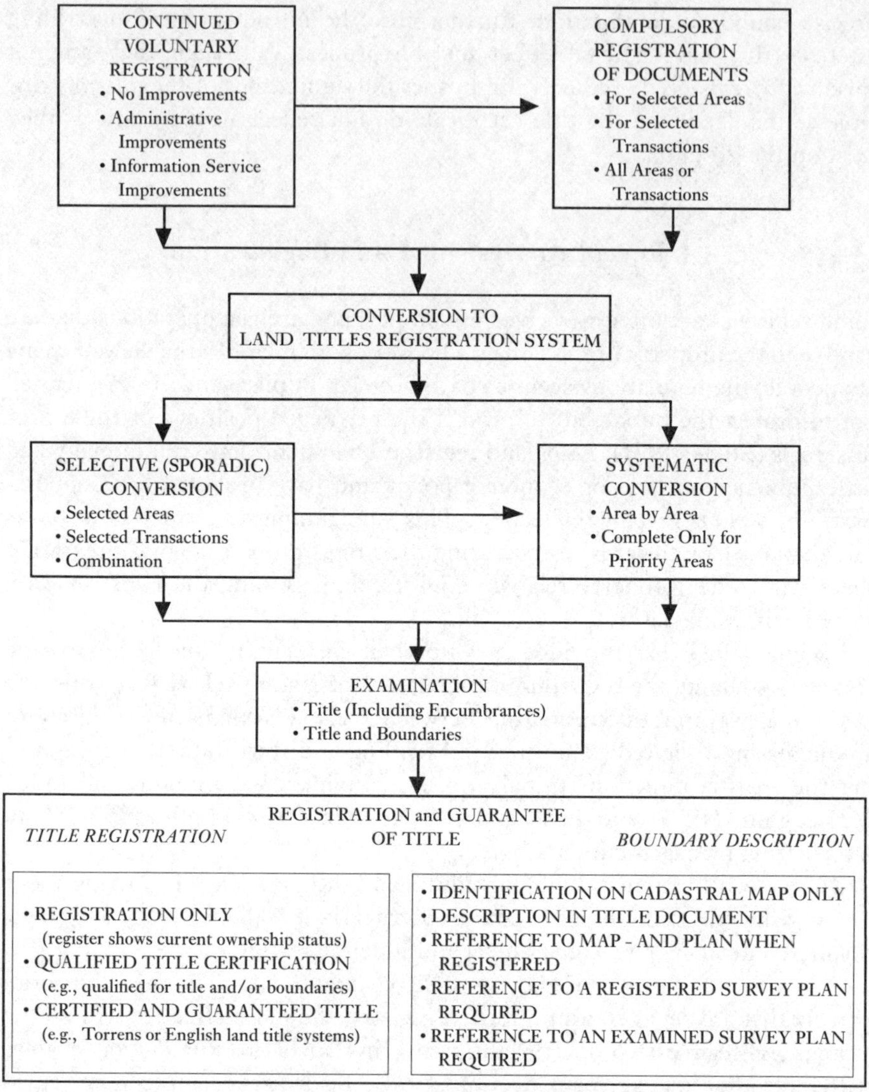

FIG. 4.1 Strengthening the registration function

movement back to informality if people feel that it costs too much to comply with the law. Many jurisdictions provide legal incentives for people to register by giving priority to a registered interest over an unregistered one. Often, however, third party requirements may count as much or even more than government sanctions and incentives. Commercial credit companies, for example, may be satisfied with the operations of a registry system and may in turn provide the incentive for potential borrowers to formalize their property.

The second of Palmer's five criteria is quality control. Since interests in land are an institutional construct, they must be represented by information; the more reliable the information, the more useful it becomes. In part, the quality of information may be controlled through examinations of incoming documents. Using terminology developed by McLaughlin (1975), an 'active' registry examines the substance of incoming documents and allows only those that pass muster to be registered, while a 'passive' registry merely files or records any incoming document without evaluating its contents.

In part, control of quality may be enforced by certification of documents by property professionals (in the public and private sector) before the documentation even reaches the registry office. Improved quality does not come only from examinations; it also requires effective techniques for storing and retrieving information. For example, a registry system employing unique permanent numbers to identify parcels will have less ambiguity than a system using a continually changing grantor–grantee index as its primary key. Good information management techniques are clearly made better when appropriate computer technology is used. However, as numerous case studies testify, applying computers to inefficient operations does little to improve the total process; the end result is often that the mess has been automated at great expense.

The third criterion is currency, which is the extent to which information in the registry system is kept up to date. For example, information about a land parcel may have been examined at a registry office and found acceptable at the time when the property was transferred, but that information is accurate only until the next transfer. Consequently, even though there is a system in which the entry of information into the registers has been examined precisely, with great care and often great expense, the records may in due course no longer reflect titles that are respected and accepted by the community. Increasingly, especially in countries with low levels of income, the title registers may show people who are dead as the registered owners of land since relatively high transaction costs discourage heirs and other subsequent owners from seeking formal recognition of transfers. There may be no uncertainty amongst those living in the community, even though the official records are incorrect. As with the issue of jurisdiction-wide coverage, merely making the registration of subsequent transactions compulsory will not necessarily provide the result that has been sought after. It is essential that the benefits of continuing registration of transactions must be seen to exceed the associated costs.

The fourth criterion, guarantee, measures the extent to which the jurisdiction stands behind its public registry. 'Torrens-based' jurisdictions guarantee their registers in two senses. First there is an affirmative warranty with the title of the registered owner being as stated on the register. This vesting provision goes beyond merely declaring an existing title to be good. It positively confers the title since without registration the owner might have no title at all. Second, there is a negative warranty inasmuch as the statutes guarantee that the title is not affected by anything not shown on the register. That is, it is not only unnecessary but also impossible to establish a right in the land by any other means (Simpson 1976).

In general, Roman law-based jurisdictions do not maintain 'positive' registers in the way that Torrens-based registration proves ownership. Nonetheless, some form of guarantee is provided since formally registered ownership may only be overruled by court order (although a registrar may correct minor errors). Typically, to void a registered ownership interest, a court must find evidence of bad faith, and the challenger has the burden of showing that the registered owner did not enter the transaction in good faith. The guarantee is much stronger in Roman-Dutch countries such as South Africa which have a negative warranty of title (similar to those of Torrens systems) in that there is no other way in which ownership of land can be acquired other than through the medium of the registry (Simpson 1976). Many jurisdictions, however, do not guarantee entries in the registry and transactions are merely recorded and not examined.

The final of Palmer's five criteria, indemnity, addresses what happens if something should go wrong with the registration system. In other words, it represents a 'comfort factor' for users of a registry system. Some jurisdictions, such as Sweden and England, may have a comprehensive indemnification programme whereby a person suffering loss of an interest because of an error in the registry receives compensation regardless of the source of the error. Others, including South Africa and Namibia, may indemnify a person who suffers loss as a result of the failure of registry officials to act in good faith or to exercise reasonable care of diligence in carrying out their duties. Many jurisdictions, however, offer no compensation and expect an injured party to recover damages from those who caused the loss in the case of fraud or from legal professionals in the case of their negligence. A private indemnification programme has emerged in the USA in the form of title insurance. Under such a programme, the insurance company will indemnify a person suffering loss as a result of errors in the registration. However, what is insured is merely the accuracy of the title search by the title insurance company and not the perfection of the title (Bernhardt 1993).

4.5 Re-engineering Land Registration Systems

In recent years, several jurisdictions have begun to re-examine the nature and function of their registration operations. This is leading to a restructuring or re-engineering of land registries, and is focused around four closely related sets of reforms—legal, judicial, administrative, and technical—as shown in Figure 4.2.

The legal reforms have addressed the modernization, standardization, and simplification of legislation relating to land and property registration. They have included:

- simplifying the nature of title that can be registered, for instance reducing this to either freehold or leasehold;
- reducing the number of overriding interests (eliminating rights that do not have a significant impact on the property);
- making registration compulsory so that economies of scale apply and quality controls can be applied through the examination of abutting properties;
- converting indexes based upon names into parcel-based indexes that undergo less frequent change;
- introducing or restricting state guarantees on titles and on boundary records, based on risk management; and
- coordinating registration law reforms with other property-related legislation, for example that associated with physical planning and land use.

Other legal reforms have included amendments to statutes of limitation (which lay down the time that must elapse for prescriptive rights and the period of notice that must accompany initial registrations); mining laws (defining how

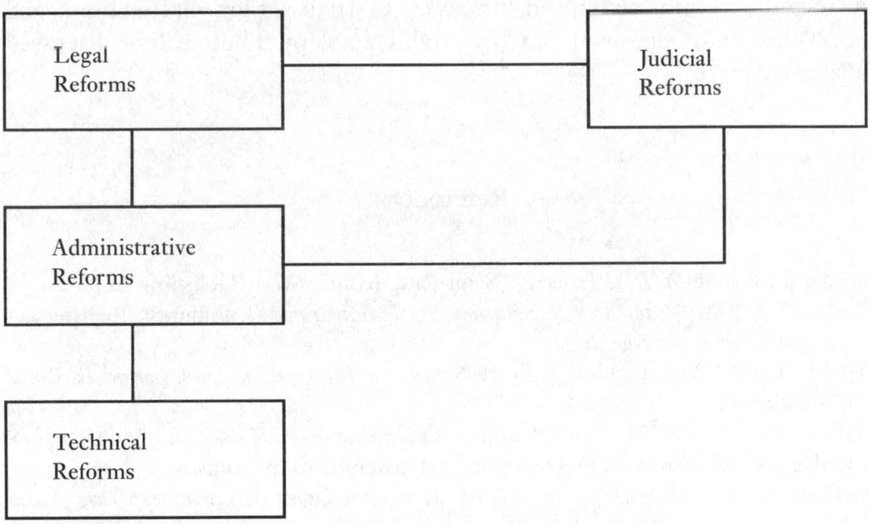

FIG. **4.2** Re-engineering of land registration system

and where mining licences and leases are to be registered); and rules of evidence (prescribing what evidence can be used to serve as a security right).

Judicial reforms have been of special importance in many developing countries and have sought to (i) set out the most appropriate role for the judiciary, for instance by giving more responsibility to administrative officers and tribunals and by clarifying the role of notaries; (ii) standardize and formalize judicial procedures; and (iii) introduce concepts and procedures for accountability. Replacing judicial processes with administrative tribunals for the resolution of property disputes has had increasing support in recent years. More generally, there has been an increasing emphasis on community-based resolution of titling disputes (Palmer 1996).

Administrative reforms have included:

- improving record management through standardization of procedures and minimizing duplication;
- introducing risk management principles in the examination and handling of documents;
- advancing new strategies and policies for the efficient and effective employment of professional and support staff, including the use of the private sector;
- developing 'one-stop shopping' facilities for the provision of public services, so that customers can obtain answers to their queries through one point of access into the system;
- decentralizing selected operations to the local community.

Technical reforms have largely been concerned with the computerization of the records, the provision of on-line access to databases for selected users, and modernization of surveying practices and technology. These will be discussed in later chapters.

References

Bernhardt, R. (1993), *Real Property* (Saint Paul, Minn.: West Publishing Co.).

Burdon, I. (1998), *Automated Registration of Title to Land* (Edinburgh: Registers of Scotland Executive Agency).

Dale, P. F., and McLaughlin, J. D. (1988), *Land Information Management* (Oxford: Clarendon Press).

McLaughlin, J. (1975), 'The Nature, Design and Function of Multipurpose Cadastres', Madison: Ph.D. Dissertation, University of Wisconsin.

Nichols, S. E. (1993), 'Land Registration: Managing Information for Land Administration', Ph.D. dissertation, Department of Surveying Engineering Technical Report No. 168, University of New Brunswick.

Palmer, D. (1996), 'Incentive-Based Maintenance of Land Registration Systems', Gainesville: Ph.D. Dissertation, University of Florida.

Simpson, S. R. (1976), *Land Law and Registration* (Cambridge: Cambridge University Press).

Stigler, G. (1971), 'The Theory of Economic Regulation', *Bell Journal of Economics and Management Science*, 2/1.

5

Surveying and Mapping

5.1 Cadastral Surveying

Recent major property formalization projects have had total estimated costs ranging from US$20 million to more than US$250 million with loans ranging from US$2 million (to support feasibility studies) up to approximately US$100 million. In general these costs have been required for:

- institution strengthening: 10–15 per cent
- mapping: 20–5 per cent
- adjudication and surveying: 30–50 per cent
- registration: 20–5 per cent.

This chapter reviews aspects of the adjudication, surveying and mapping of property, processes that are used to identify and record the land to which ownership rights are attached. There are essentially four basic processes involved:

- Adjudication and the definition of the boundaries of the parcel.
- Legal delineation and monumentation of property boundaries.
- Surveying all the relevant data.
- Preparation of descriptions of the properties, usually in map form.

These processes are essential components in the initial formalization of property and in the subsequent use and transfer of formal property rights. They provide much of the information necessary for the overall planning and coordination of land titling projects, for assessing evidence in support of the initial registration of title, for the actual delimitation and registration of boundaries. They also form the basis for the subsequent re-establishment of boundaries in the case of uncertainty, dispute, or subdivision.

Cadastral surveying is the term generally used to describe the gathering and recording of data about land parcels, even where the records do not form part of an official cadastre. When properties are initially registered, government officials have traditionally undertaken the processes of cadastral surveying and land title adjudication. The costs are then paid for by the state or through aid programmes. More recently, private sector surveyors have carried out some of the work on contract in an attempt to reduce actual costs and to accelerate the pace of projects. The surveys of changes to the boundaries that take place after the initial survey (through consolidation, subdivision, amendment, etc.) may be undertaken by the public sector but are more often carried out by private licensed surveyors.

Cadastral surveys are concerned with geometrical data, especially the size, shape, and location of each land parcel. In some jurisdictions, cadastral surveying is solely concerned with the location of property boundaries while in others it includes all things attached to the soil. The latter concept encompasses both land and its associated buildings, objects that in some jurisdictions are treated as totally separate entities.

As noted earlier, the land or cadastral parcel is defined in theory as a continuous area, or more appropriately volume, that is identified by a unique set of homogeneous property rights. However, in practice this definition gives rise to difficulties in the case of:

- Non-continuous areas—for instance where a farmer owns non-adjacent fields or a residential property includes an area divided by a road that is in the ownership of the local authority.
- Administrative boundaries that may divide a property between two authorities so that although in one ownership, development approval must be sought from a different agency depending on which area of the property is to be developed.
- Changes in natural features where the ownership may be homogeneous but the land cannot be used homogeneously—for example where parts of a site may be suitable for building while others may be better used for agriculture or access.
- Differences between tenure and use, where for instance parts of a property are subject to leasehold or there are specific rights of way across only certain areas of the property.
- Patchworks of limited interests in the land—in Slovakia for example the restitution process has resulted in cases where a single field being worked as an agricultural unit has over 300 owers and 1,000 co-owners.

5.2 Adjudication

The function of property adjudication is to resolve disputes and uncertainties pertaining to who owns what property. It may focus solely on problems that exist when property is first formalized but in some jurisdictions it is also involved in the many problems that arise after formalization. Initial adjudication poses the same basic challenge as other aspects of the titling and registration process: how to accomplish it quickly and efficiently, with as much community-based support and involvement as possible, while maintaining the minimum level of security and professional support necessary to achieve the goals.

Adjudication is the first step in the registration of title to land and encompasses procedures for determining what rights exist on the ground. In theory, adjudication does not alter these rights but rather confirms the existing legal position. The procedures followed must be based on due processes of the law so that the resulting decisions determine definitively who owns the land. In some cases, for instance where the evidence is uncertain, it may be necessary to register the title as a Qualified or Possessory Title and in such cases time may be required to cure the weakness. Thus registration can go ahead on the basis of the best evidence available. Then if after a set period of time, such as ten or fifteen years, no one has come forward to challenge the title, the registration can automatically become definitive. What was a qualified or possessory title would then become a full title and be guaranteed by the State.

In many jurisdictions, there are possessory claims to land that need to be resolved. Under the common law doctrine of adverse possession, the peaceful and unchallenged occupation of land can lead to the prescription of full rights so that after a specified period of time the peaceful occupier can claim title to the land. Under normal circumstances, no one can inherit a title better than that of his or her predecessor but landowners can refuse to pass on all the rights that they possess. Thus, a landowner can impose restrictions on the future use of the land at the time of sale by, for example, specifying that the land cannot be used for a specific purpose such as certain forms of commercial enterprise. By this means, existing rights can be reduced on the transfer of land but they cannot be increased. The only way that they can grow is by prescription, through the lapse of time.

In some cases information on who owns the land may already exist in deeds registries and other archives. This may happen when, for example, a deeds registration system is being upgraded to title registration. In such cases the procedures for adjudication begin by identifying where the data are located and then checking their veracity before registering the title. It is however generally wise to check on the ground to ensure that the Deeds Registry records are up to date and that there have been no unrecorded changes to the property or that it has now passed through inheritance to an heir.

Title registration can also start with the examination of existing unregistered deeds and other documents, a process that in England and Wales is known as first registration. All relevant documents relating to past ownership are submitted, usually by solicitors acting on the behalf of their client, to the Land Registry where a decision is taken as to what land is involved, where approximately its boundaries lie, and the nature and limitations on the tenure. No automatic check is made on the ground although sometimes this is necessary if there is conflicting evidence. Relevant information is then entered on the Certificate of Title, the definitive copy of which is formally held in the Registry.

In many of the countries in east and central Europe there has been a process of land restitution in which the old pattern of landownership that existed before the communist era has been re-established. Records of landownership held in the national archives have been scrutinized and, subject to subsequent inheritance of the rights and the feasibility of restoring the old property boundaries, have been used to rebuild the old land tenure system.

In other areas of the world where no formal records of ownership have ever existed, different procedures are followed and the local community must be consulted. The determination of who owns what land may be undertaken on a sporadic or systematic basis as discussed in §3.3. In order to carry out systematic adjudication it is necessary to:

1. publish a law to control the proceedings;
2. select optimum areas in which to operate, priority being given to areas where:
   - there is a significant level of dealing in land
   - there is need for security for credit
   - there is a high incidence of litigation
   - changes in land use/holdings are proposed
   - lack of certainty inhibits development and, in practice,
   - it is politically expedient;
3. formally announce the areas affected, giving the proposals maximum publicity;
4. appoint adjudication, demarcation, survey, and recording or registration officers;
5. conduct the adjudication on the ground;
6. allow time for objections, appeals, and the rectification of the initial adjudication where appropriate; and
7. finally enter the details on the register.

The processes may also be compulsory or voluntary. Where land reform, and in particular land consolidation, is planned as part of the land titling process, registration should be compulsory such that all interests are identified. Voluntary registration allows for the process to be market driven but does not allow for the comprehensive collection of land records that may be beneficial to the community as a whole. It also runs the danger of encouraging the registration of only weak titles where conveyancers are unwilling to provide a warranty for a transaction; good titles then tend to become those that have not been registered.

5.3 Boundaries

Boundaries mark the limit of each tract of real estate. In legal terms, a boundary is an imaginary line that divides two adjoining estates while in common language the term denotes the physical objects by reference to which this line of division is described. Thus a fence or hedge may be described as a boundary although from a legal perspective such physical objects may not coincide with the legal limits of a property. In law, fences and hedges are regarded as guards against intrusion and not necessarily indications of the position of a legal boundary.

A monument is any tangible landmark indicating boundaries. Monumentation may be in the form of linear features such as walls, fences, hedges, or streams or in point form such as the concrete beacon. A monument should be permanent, stable, certain of identity, and preferably visible. The timing of monumentation for cadastral purposes is often critical to its permanence since if monuments are emplaced prior to development they may be damaged, destroyed, or moved during the construction process.

Boundaries may be specific (sometimes referred to as 'fixed'), in which case the precise line of the boundary can be determined. Alternatively they may be general, in which case the official register only shows the approximate line of the boundary, precise details of which can only be established by further investigation on the ground. There are three categories of specific boundary, namely those that are:

1. defined on the ground prior to development and identified, for example, in documents of sale;
2. identified after development, for example when the line of the boundary is agreed between neighbours at the time of adjudication; or
3. defined by surveys to specified standards.

Each type reflects the fact that the precise position of a boundary has been determined.

There are also three categories of general boundary, namely those where:

1. the ownership of the boundary feature is not established, so that the boundary may be one side of a hedge or the other or down the middle;
2. the boundary is the indeterminate edge of a natural feature such as a forest; and
3. the position of any boundary is regarded as approximate so that the register may be kept free from boundary disputes.

Each concept reflects the fact that the precise line of the boundary has not been adjudicated on the ground.

The accuracy of survey measurements in specific boundary systems is of secondary importance; monumentation and defined limits of possession on the ground are the primary considerations. Any preference between specific and general boundaries should depend on whether in a developed area it is economic to carry out adjudication systematically or whether the approach should be sporadic based upon demand by landowners. It should also depend on finding the best method for keeping the register as a reflection of the ground.

Boundaries are not necessarily permanent. They may be varied by:

- Agreement between neighbours, for instance when a fence falls down and the adjoining parties agree to re-erect it along a slightly different line.
- Adverse possession and the peaceful occupation of the land so that the accepted use of a strip of land can lead to change in the land's formal legal status.
- Estoppel, a legal process for barring other claims, so that by default an original owner cannot reclaim land over which a neighbour has built.
- Rectification by officially correcting the register subject to the agreement of the parties concerned or as a result of a court order.
- Registration by official decision of the Registrar especially at the time of first registration.
- Natural forces such as accretion and avulsion in which a boundary defined by a river may change as the river alters its course; or
- Ruling of the Courts in the case of disputes.

Boundaries are a matter of law; the role of the land surveyor is to collect factual evidence to assist in their description. Evidence for the location of boundaries may come from:

- Title deeds, that is the formal documents describing the title.
- Witnesses who are prepared legally to swear to what they know.
- Public documents, for instance containing details of official administrative boundaries.
- Ancient documents and old maps.
- Declarations of deceased persons, for instance in their Wills.
- Acts of ownership and the way in which the land has been occupied and used.
- Modes of construction, such as the way that a fence has been built and maintained.
- Local custom and the way that the land has traditionally been used, especially land belonging to the community.

In seeking to determine the location of boundaries from old descriptions there are certain assumptions normally made regarding the interpretation of certain words recited in the deeds and other legal documents. Expressions describing plans as 'for identification purposes only' may still be taken as evidence even though they imply that they are only sketches. When a property description refers to a plan, that plan becomes part of the description and what is written on it has legal significance. Written information in verbal form tends to be superior to that given in numerical form—thus if a dimension is written in words as 'ten metres' but appears elsewhere written as '1.00 m' then unless other information is available, the written figures will prevail. When a conveyance is reduced to writing, it is assumed to contain all the terms intended and is construed, where possible, to give effect to all the terms.

That having been said, when restoring boundaries, the surveyor is required to retrace the actual lines on the ground that were laid out by the original surveyor rather than the lines indicated by field records. To achieve this, measurement is only part of the process. The Courts normally regard measurements as being inferior to monuments if conflict of evidence exists—as the saying goes, 'Pegs are paramount to plans: marks come before measurements.' While this of course depends upon the jurisdiction concerned, in general the law is less concerned with precise measurements and more concerned with physical features than are surveyors.

5.4 Survey Methods

Surveys of properties and associated boundaries are required to determine their locations and to provide evidence for their future retracement. A survey may be needed in the initial setting out of land parcels, in recording what has already been built, and in re-establishing boundaries either in the case of dispute or where subdivision is to take place. Surveying is essentially an investment in the future to ensure the long-term maintenance of the parcellation of the land. The costs of cadastral surveying depend very much on the precision that is sought—that is how precisely the boundaries are defined and must be measured.

One problem for the land surveyor is therefore to determine what are the optimum standards, within either a specific or a general boundary system, that will in the long term produce significantly greater benefits and lead to better overall land administration. Surveys should be accurate but the level of precision with which the measurements should be recorded will vary depending on local circumstances.

In many countries, the techniques that must be used in cadastral surveying are prescribed in the law and in regulations that specify the standards that are to be achieved and the methods that must be used to deliver them. Surveyors may also need to be licensed to carry out their work. Regulations and processes for licensing surveyors were introduced in many countries to ensure that the right quality of data was produced. These standards were and in many cases still are monitored by the central government cadastral mapping agency which then takes responsibility for the accuracy of the work.

The choice of the most appropriate method of survey is dictated either by regulation, the type of monumentation that is used to demarcate boundaries, the type of terrain, the needs of quality control, or the form of end product that has been specified. The choice should also depend on cost and the levels of risk involved although these factors are often ignored in the selection process. Where surveys are carried out by government agencies, true cost figures are often not available—thus for example the actual costs for overheads may be uncertain or be ignored. Similarly, the level of risk is rarely quantified and obtaining hard rather than anecdotal evidence on the number of boundary disputes or the possible damaging effect of lower order accuracy mapping is often difficult.

In practice, the cost of survey can be almost half the total cost of registering many titles. In a number of central and eastern European countries there are aspirations to resurvey the old cadastral boundaries to new standards of accuracy using modern technology; in the case of Poland, for example, this cost is estimated to exceed US$ 1 billion. There is neither the time nor the capital available to implement such a strategy; hence alternative approaches must be adopted. According to Baldwin et al. (1998), the costs in millions of ECU (approximately equivalent to the US Dollar) are likely to be as shown in Table 5.1.

TABLE **5.1** Costs of high-and medium-accuracy cadastral mapping in millions of ECU (from Baldwin et al. 1998)

	Time until completion (years)	Cost of high-accuracy cadastral mapping (million ECU)[a]	Cost of medium-accuracy cadastral mapping (million ECU)[a]
Establishing basic technical infrastructure	5–10	35	35
Loading/verifying title data	5–7	25	25
Loading cadastral mapping (high accuracy)	15	1,500	
Cadastral map data conversion (medium accuracy)	7–10		75
Operational costs at 25 million ECU per year	10	250	250
Reinvestments/renewal/maintenance at 10% per year	10	35	35
Value added services development	5–10	25	25
Total costs for 10-year period	10	1,870	445

[a] Costs for mid-size central European country with: population of 10 million, area of 100,000 sq.km, 75,000 cadastral map sheets.

Source: Baldwin et al. 1998.

There are two broad categories of surveying technique—field survey and photogrammetry. There are also two categories of output—graphical and digital. The traditional approach to cadastral surveying is through the use of ground survey techniques with a surveyor recording data either in a field notebook or in more recent times electronically. An alternative is to use photogrammetric techniques using aerial photography in which case most of the measurements are recorded in an office. Both sets of techniques are capable of producing the necessary standards of accuracy and precision for most land administration purposes. Photogrammetry cannot however be used for setting out, other than as a basis for the design of plot layouts, and its use is dependent on the existence of property boundaries that are visible from the air.

Similar constraints exist with the use of satellite imagery and remote sensing techniques. In addition, they do not in general meet the standards of accuracy and precision needed for detailed cadastral survey. As the technology improves and very large-scale images can be obtained using satellites, things may change but for the present satellite imagery remains too crude for extended use.

In new areas where titles to the land are to be brought under formal control, photogrammetric techniques offer a relatively rapid, cost effective, mass production way to achieve first registration. The unit cost of surveying each land parcel can be kept low through economies of scale. Graphic standards of accuracy, in which the precision of measurement is determined by the thickness of a line drawn on the map, are easy to achieve. Similarly, precise coordinate values can be obtained, given the appropriate technology and human resources. The issues are those of cost and effectiveness under the circumstances, and the legitimacy of such techniques under regulations that have been designed for surveyors working on the ground.

One advantage of aerial methods is that the photography on which such measurements are based provides a historical record of the landscape that can be revisited in the future to see what changes have taken place and even to re-measure conditions in the past. Thus where disputes arise over whether a boundary has been moved, old aerial photographs can provide crucial evidence.

Most cadastral surveys are however undertaken using theodolites, steel tapes, or electronic distance measuring (EDM) systems, including 'total stations'. Increasing use is being made of global positioning systems (GPS) techniques that can provide coordinate values for points on the ground to a high level of precision, for instance to better than one centimetre. Such high-precision work takes time and requires relatively expensive equipment although prices continue to decline. GPS techniques can also be used in support of photogrammetric surveys. As the price of GPS instrumentation falls so does the unit cost of fixing any individual boundary point.

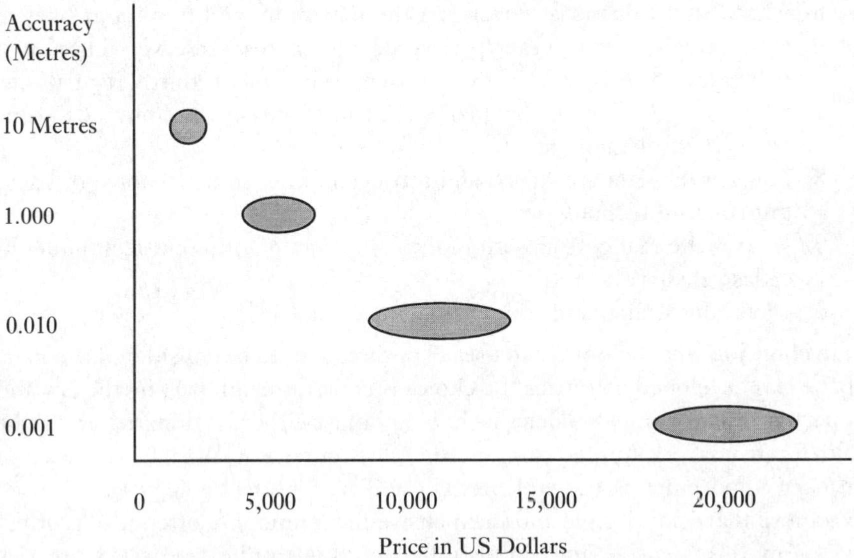

FIG. 5.1 Price of global position system technology

The traditional approach to boundary surveying has focused on the measurement of the bearings and distances between adjoining boundary beacons. Technology such as EDM and GPS makes it cost effective to measure the coordinates of boundary points, values that can easily be handled by computers.

In accordance with the UN Land Administration Guidelines (UNECE 1996), the selection of the most appropriate techniques for cadastral survey and mapping depends, amongst other things, on:

1. what physical boundary features are used to delimit property—for instance whether there are fences or hedges or whether only the boundary line turning points are monumented;
2. whether there are cultural or legal restrictions affecting the choice of boundary delimiter; for instance in some communities physical barriers to entry are considered anti-social while some residential areas of modern towns are 'open planned';
3. whether the boundary points or lines are visible or can be made visible on aerial photographs so that photogrammetric techniques can be used;
4. what is the legal status of maps and plans kept within the land registration system;
5. what are the legal requirements for land parcels to be surveyed, what accuracy standards are laid down, and whether the field notes of surveyors are treated as legal evidence;

6. what restrictions exist governing the maximum and minimum sizes of land parcels, and whether precise details on road reserve widths, plot frontages, etc. form part of the building development control regulations;
7. who is responsible for the quality control of surveys and how such controls are implemented;
8. how survey data are stored and retrieved either in paper form or using information technology;
9. what is the real cost in both money and time of different techniques of cadastral surveys; and
10. what educational and training skills are available.

In choosing what technique to use, all options should be considered. In many of the less developed countries the choice is constrained in part by the law and in part by finance since modern technology must either be donated or else be paid for from very limited amounts of hard currency. What is however of primary importance is the end product and not the technology that is used to achieve that end. Simple and inexpensive techniques are often just as effective as more expensive and sophisticated methods. The key issues are risk management and cost.

5.5 Property Descriptions and Referencing

A register of titles normally contains two elements—a record of the attributes associated with each parcel and a description of the land to which these attributes refer. The attributes that may be recorded in the registers include

- the names of the owners;
- the nature of the tenure (such as freehold or leasehold);
- the price paid for the land on transfer;
- any restrictions on the use of the parcel such as restrictive covenants or charges where the land is subject to mortgage;
- any exclusion of rights to minerals below the soil; and
- any caveats or cautions that may require a third party to be informed if any dealings in the land are proposed.

Depending on the legal system, certain categories of overriding interests may be assumed to apply even though they are included within the entries on the registers—such as the rights of others to light or the restrictions imposed on all properties by Town and Country Planning legislation. Such overriding interests are restrictions that legally bind the landowner and affect the use of the land and the obligations of the landowner.

Additional information of a multi-purpose nature, such as the land use or the soil type, may also be included. In many cases, for example where a right of way exists over certain parts of the land, there may be evidence on an attached plan that supplements what is written on the register. This plan then would become a formal part of the description.

The description of each parcel of land is usually based on a plan and is linked to either verbal or numerical data. The function of a cadastral plan is first and foremost to identify the parcel of land that has been referred to in the written parts of the registers. This may of course be achieved by a simple parcel reference number relating to the house and street and town in which the property is located, such as '23, High Street, Newtown'. Such a verbal description would identify the parcel, whose limits would then be determined by inspection on the ground.

Frequently however this is insufficient; hence the parcel description should provide additional information such as the shape and size of the property and at least the approximate location of its boundaries. It should also show the relative location of the property by providing some details about adjoining parcels. The description should enable the boundaries to be relocated in cases of dispute or uncertainty, enable subdivision to take place within or up to the limit of the existing plot, and permit areas to be calculated for the purposes of planning or assessment. In addition, it should form the basis for land information management and the multi-purpose cadastre.

Parcel descriptions may be verbal—for example 'all that land between the river Exe and the main road between Newtown and Wyemouth'; numerical (including the coordinates of or the bearings and distances between corner monuments); or graphical with the information plotted on a plan. Verbal descriptions on their own are inadequate for most purposes. An exception is where they refer to the bounds or abuttals, that is the details of what adjoins the property—for example 'up to the edge of the land owned by Mr Farmer'. Such descriptions reflect the importance of occupation and use of the land and may help in determining boundaries on the ground. In general, however, they are significantly inferior to the other methods and produce little information of value for other applications.

Graphical methods are the most effective way to indicate boundaries and are easy to comprehend. They allow most relevant information to be communicated quickly and cheaply. Cadastral plans should be produced at a scale that is large enough for every separate parcel to appear as a recognizable unit, bearing in mind that the larger the scale, the more expensive in general will be the mapping. The plan should show not only the boundary monuments such as pegs but also relevant topographical features such as buildings and fences for they will help in the identification of the boundaries. In urban areas, a scale of 1 : 1,000 is almost always sufficient—there are of course exceptions but even in the centre of a city such as London where land values are amongst the highest in the world, a scale of 1 : 1,250 is often sufficient. In rural areas a scale of 1:5,000, 1:10,000, or even 1:20,000 may be adequate depending on the size of the parcels and the manner of their demarcation.

Property descriptions are increasingly based on numerical information. Modern field survey techniques normally produce numerical data whilst with the increasing use of computers, even graphical information is being converted into digital form. From such data it is relatively easy to produce the description of property boundaries either in terms of coordinates or by reference to the metes, that is the bearings and distances between adjoining points. In some jurisdictions the metes are combined with the bounds but in such cases it is normal for the bounds to take precedence if there is any conflict of evidence. The numerical data may be held within a computer or on printed lists.

The relative weight given to different types of evidence varies between jurisdictions. In most countries the marks found in the ground take precedence over the numerical values. Hence if a peg is found at a measurement that differs from that shown on the record then, unless there is other evidence to show that the peg has been moved, its present position will be accepted. Historically, this has been in part because the quality of some survey measurements was not good, but more particularly because pegs generally represented the limits of possession and occupation of the land. Measurements were more open to dispute.

Increasingly, many land surveyors argue that the quality of survey measurement is now such that greater certainty can come from trusting the record, as in the English fixed boundary approach, rather than the pegs on the ground which are at all times subject to movement. This argument is reinforced by the confidence that many people have in the quality of data held in numerical form. On balance, however, the prevailing sentiment still holds that it is the monuments that should matter most.

The various methods for describing the limits of each parcel must include an element that uniquely refers to the parcel as a whole. Even with a graph-ical system it is necessary to have a parcel reference number that identifies the parcel and allows for cross-referencing within the registers and any other filing systems. Given the parcel reference number, it should then be possible to access all the details of the parcel including its attributes and its spatial description. A variety of referencing systems have been adopted, including the names of people involved in the transactions, title reference numbers, property names and street addresses, and grid coordinates or geo-codes.

The important issues are to get consistency between different agencies addressing the same parcel and to have a system that is:

1. easy to understand, for thereby less confusion and mistakes are likely;
2. easy to remember so that landowners can recall their parcel references;
3. easy to use both by the general public and administrators and in computers;
4. permanent so that for instance the parcel reference does not change with the sale of the property;
5. capable of being updated when there is a subdivision or amalgamation of two adjoining properties;
6. unique so that no two parcels have the same reference and there is a one-to-one correspondence between what is on the ground and what is referred to in the registers;
7. accurate and unlikely to be transcribed in error;
8. flexible so that it can be used for a variety of purposes from registration of title through to all forms of land administration; and
9. economic to introduce and to maintain.

Unique property identifiers are an essential element in an integrated land information system. Indeed, given well-monumented boundaries, an adjudication process that identifies the owner of the land delineated and a unique parcel reference number then the benefits of precise cadastral surveys only arise in the longer term. There can be important efficiency gains through computerizing the text data in an existing land registration system as for example has been the experience in Sweden. The added value that comes from good mapping, especially in digital form, may be relatively small although significant. Where parcels are not well defined on the ground, the benefits of mapping are very much greater.

References

Baldwin, R. A., Bogaerts, T., Csaki, C., Dale, P. F., and Zichy, A. (1998), *The Development of Land Markets in Central and Eastern Europe*, Report to the European Union Action for Cooperation in the Field of Economics.

United Nations Economic Commission for Europe (UNECE) (1996), *Land Administration Guidelines: With Special Reference to Countries in Transition* (New York and Geneva: United Nations).

6

Land Values and the Fiscal Cadastre

6.1 Land as a Financial Asset

Classic economic theory holds that there are three vehicles for generating wealth in an economy—capital, labour, and land. Land is fundamental, for labour cannot live without space and capital cannot be managed without offices and the infrastructure that is built upon the land. The management of land has social, political, and economic dimensions. While the post-war land reforms were driven largely by political agendas, current reforms are primarily concerned with the development of land markets.

In their study of urban land markets, the Organization for Economic Cooperation and Development (OECD) pointed out that:

> Land plays an important role as a financial asset. It is an important element in the portfolios of central and local government, nationalized industries, private companies and financial institutions. Financial markets and property markets are intimately connected. Land, especially seen from an historical perspective, is often considered from an investor's point of view as a superior asset to the financial assets available on capital markets, mainly because of the potential of land to maintain its value over time and because of favourable tax treatment. The more capital and land markets are developed, the higher is the degree of possible substitution between land and other assets. Land and building values together can account for a substantial share of the market capitalization of many businesses and are often a prime consideration of corporate strategy. Stock market growth can be fuelled by rising prices in real estate markets when land is used as collateral for loans. Should land and prices fall in a volatile market place, a high level of dependency on land and property-based assets may carry the risk of serious financial disruptions.

The report went on to state that:

> Land policy cannot be effectively designed and pursued if governments do not understand how their land markets operate (OECD 1992).

Land and property are important components in any market driven economy—their value is a measure of the wealth of any society and probably accounts for more than 20 per cent of GDP (UNECE 1996). In most countries, the biggest landowner is the state. Logically, therefore, the state should manage its assets effectively and efficiently, yet many central and local governments have inadequate records of what assets they hold. Anecdotal evidence reveals many examples of municipalities, for instance, buying land that they already own or seeking to build on land that they subsequently find that they do not own. This is symptomatic of the fact that many governments fragment the processes of land management with separate policies being developed by different ministries and agencies leading to duplication, contrary policies, and confusion. Even the definition of land (whether the term, for example, includes buildings) may differ between ministries, a contradiction that has even existed, for example, in the land legislation of Latvia.

Although the municipalities normally have responsibility for the detailed control of land use, they have less direct control over taxing land and the creation of local land banks. Land taxation is a mechanism that can be used to reduce demand for land in areas of overdevelopment or stimulate demand, for instance by encouraging vacant or underutilized land to be brought onto the market. Similarly, if municipalities have reserve stocks of land, they can use these local land banks to regulate land supply, releasing land onto the market to stimulate development or holding it back to damp down demand. Furthermore, when urban expansion takes place, the municipalities can use their reserves of land for the provision of services such as schools and other municipal facilities. Without such a resource, they can be forced to buy land for civic purposes at prices that are at the top end of the market when development that they have helped to bring about is nearing completion.

Municipalities often do not have the freedom to operate in such a way, either because they lack the financial resources or are constrained by the law. Also, municipalities implement central government policies and it is at that level especially that there is often a failure to understand the importance of land and the concomitant need to:

- monitor land availability, such that the authorities can influence the demand for land (through taxation or regulation) and its supply for development;
- develop land use policies that optimize available resources;
- monitor land and property transactions so that if the authorities want to interfere in the land market they can do so at acceptable prices, or tax land at a fair rate;
- tax land through systems that are simple and that generate significant revenue in an equitable manner;
- harmonize land tax policies with land use policies;
- consider the creation of land banks that can influence local land markets;
- lease government assets at market rates;

- monitor land and property assets so that those surplus to requirement can be released.

6.2 Land Taxation

Economists have promoted the virtues of land taxation for centuries. Advocates from George (1879) to Netzer (1998) have argued that, unlike other taxes, a tax on land neither distorts economic decision-making nor lowers the efficiency of using market forces to allocate resources. Others have been less enthusiastic in their support (Roakes 1996). In any event, land and property taxes have a number of advantages, both in terms of providing revenues to government (especially local government) and as a tool for guiding land use and development.

Any taxation system needs to be seen to be fair and to serve social objectives that are understood and accepted by those who pay the taxes. It should raise significant revenue by an amount that is substantially in excess of the cost of its collection. An exception to this may occur when the objective of the tax is regulatory (as in taxing land in order to bring it into better use), or it is used to encourage certain forms of investment. Thus vacant or otherwise unproductive land may be taxed with the specific objective of forcing the owner either to use the land more effectively or to sell the land to those who will. The converse of this is that taxes on land and property are likely to bring about changes in land use; such changes should be anticipated, although frequently they are not.

Taxing land has the significant advantage that what is taxed can easily be identified. In most societies, many people seek either to avoid or to evade the payment of taxes. Tax avoidance is the legal process that allows citizens to make profits that are not subject to tax, by for instance placing their investments in trust funds, often in foreign countries where no tax is payable. Tax evasion, on the other hand, is a deceitful process where, for instance, a citizen does not declare to the tax authorities all sources of income such as earnings through casual labour or so-called 'black market' deals.

Taxes on land and property take two basic forms—an annual levy based on some estimate of the value of the land or property, or a levy on their transfer. Some countries use both approaches. The annual levy may be based on the estimated market value for which the property would sell under normal circumstances, or the assessed rental value of the land or property, or in some countries on the cadastral value. The latter may be calculated on the basis of a number of parameters such as the area, soil type, distance from markets, etc. A number of eastern European countries are introducing taxes on this basis since, at present, the market data are insufficient and unreliable.

The levy on land transfer, sometimes called Stamp Duty, is normally based on a scale of fees that relates to the value of the land or property being transferred. The tax should be paid every time the transfer takes place, the amount being dependent on the value of the transfer. Not all citizens of course are honest when declaring the price paid for property. In some cases there may be separate arrangements for the sale of fixtures such as curtains and carpets so that these are not formally included in the price of a house sale; this is done so that the declared value of the house is lower, thus reducing the tax burden. It is essential that government authorities keep adequate records of all transactions so that significant distortions in the declared market prices can be detected and any tax evasions investigated.

Land and property transfer taxes are in general relatively simple and easy to collect and are readily understood by taxpayers. Also, at a time when significant sums of money are changing hands between purchaser and vendor, such taxes are less painful than annual demands for payment. In addition, special taxes can be introduced to recover for the community some of the windfall profits that may arise from land use planning decisions. Thus, a farmer owning land on the edge of a town may receive large sums of money from the sale of that land if permission is granted to change its use from agriculture (where for example it may be worth $2,000 per hectare) to urban development (where it may become worth $100,000 per hectare). This enhanced value is sometimes known as betterment. A percentage of this profit can be recovered through betterment taxes and returned to the community that created it.

While high rates of tax can discourage landowners from bringing their land onto the market, failure to tax land can also have adverse consequences. In Karachi, Pakistan, for example, there are several million people who live at very high densities in squatter settlements on government-owned land. They live in mud huts (known as katchi abadis) while close by there are several hundred thousand vacant sites that have road access, water, and electricity supply but no buildings. The poor are prevented from squatting on this land and the rich are neither willing to give them access to it nor to build on it since each site represents a significant capital investment that is steadily increasing in value. Where there is a demand for land that is in scarce supply, potential sites for development increase in value; in Karachi where this has happened, the value of serviced plots that are waiting to be built on has increased faster than the rate of inflation. The policy of the government has been not to tax such land, thus making the investment in 'site and service' plots even more attractive. As a result, land that could be used to house the poor is lying unoccupied while low-income communities are forced to squat on marginal lands where there are no services.

Brazil provides an example of a country that has employed land taxes (the Imposto Territorial Rural) in an effort to stimulate rural development. The tax was implemented in the early 1960s during a period of strong pressure for land reform and was intended to encourage economic growth by relating tax concessions to levels of production. While the taxation scheme has had only limited success to date, recent research has suggested ways that the original objectives might be achieved (de Almeida and Uhl 1995).

Taxes levied annually on land and property are in essence taxes on wealth. They are often used to fund local government activities. Some critics consider that property taxes are not necessarily related to the taxpayer's ability to pay since the occupation of a property does not necessarily relate to the income that is derived from it. Some consider that the tax has a depressing effect on incentives and productivity. In the early days of land reform in countries in economic transition such as Bulgaria, it was explicitly stated that for a defined period there would be no land taxes; this was to ensure that farmers would not be discouraged from registering their titles to land.

Most countries in economic transition are now reviewing their attitudes to land and property taxes. In many of the former communist countries, the market value of property has been difficult to establish because the land market is only just begining to operate in a formal way. In Latvia for example land was until recently zoned and land taxes raised on the basis of location and land use. In 1998 however, the laws that treated land and buildings separately were changed and a new real estate taxation law was passed that will introduce new taxes in the year 2000. Under this law the taxpayers will be either individuals or legal entities who are owners or leaseholders of land or buildings. The basis for the tax will be the cadastral value, which in turn will be based on market values. Since many properties have not been sold since the introduction of land and property reforms, the actual market value must be estimated. A mass appraisal of all properties in the country is planned for completion prior to the year 2000. The situation is similar in many of the other central and eastern European countries.

Whereas property taxes are essentially concerned with improvements to the land, land taxes may be based either on the estimated market value of the land or its potential productive capacity. Agricultural land, for example, can be taxed on the estimated yield from the land per hectare multiplied by its measured area. Estimates of yield can be obtained from past experience, or from aerial photography or remotely sensed imagery that has been recorded at an appropriate time of year.

Land taxes based upon estimates of productive capacity have at one time or another been imposed in Australia, New Zealand, Denmark, South Africa and East Africa, Jamaica, Barbados, Hawaii, and western Canada. In New Zealand, the taxation of land was introduced in 1878, with the aim of breaking up the large individual holdings of agricultural landowners. Many countries still retain some variant of a land tax, though most now tax both the land and the improvements to it. Less developed countries tend to tax agricultural land while developed countries often concentrate on the taxation of urban areas.

In the case of the European Union, there is in effect a negative tax system under the Common Agricultural Policy (CAP). Subsidies are paid on the estimates of yield, the data for which is managed through an Integrated Administrative Control System (IACS). IACS is concerned both with things attached to the soil (such as crops and trees) as well as livestock (cows, sheep, pigs, etc.).

The yield that is possible from the land depends on its quality and suitability for particular purposes. Land quality refers to the condition or health of the land and its capacity to sustain any particular form of use. Common problems with land quality include land degradation, soil erosion, salination or other forms of soil quality deterioration, pollution, and the lack or excess of water. A variety of measures to determine the quality of land have been developed (see for instance Pieri et al. 1995). In the countries of east and central Europe during their periods under communist rule, detailed records were maintained of land quality with complex calculations from the data as to their suitability for different types of crops. Many of the data were not used as in practice it is the people on the ground who are often better judges of what should be grown. The data have more recently been used to determine the value rather than the quality of land, estimates that have been at wide variance with what the market is actually prepared to pay. Both overestimation and underestimation of the true market value have been common, the former resulting in an informal market, the latter in substantial profit for citizens purchasing the land.

6.3 Land Values

Land quality in the sense of its fitness for a specific purpose can be measured on the basis of its physical qualities such as its soils, aspect, rainfall expectancy, and location. Such characteristics contribute to but do not determine what the land will be worth in an open market. Land value, in a market sense, is not the same as land quality. People pay for land what they perceive that it is worth to them. Land values are determined by a variety of factors including the quality of the land, legal constraints and the intended use of the land, and the general state of the local economy. The actual price that is paid in any transaction is what a purchaser is willing to pay and what the seller is willing to accept at the time. The 'price paid' may or may not represent what might be normal within the market since for any given transaction the vendor may be willing to accept a lower price because of his or her urgent need to sell. Alternatively, the purchaser may pay more than would otherwise have been expected because of personal circumstances.

The term 'market price' is used to mean what, when all other things are equal, a property would expect to be sold for at a given time—what the Royal Institution of Chartered Surveyors defines as 'The estimated amount for which an asset should exchange at the date of valuation between a willing buyer and a willing seller in an arms length transaction after proper marketing wherein the parties had each acted knowledgeably, prudently and without compulsion' (RICS 1996).

Time is important since, due to any number of external factors, the actual market may fluctuate depending on such factors as the cost of borrowing money and the current interest rates, the rate of inflation, and government tax policies.

The cost to construct a property may differ significantly from its market price. There are direct and indirect costs in constructing a building. Direct costs include labour rates, the price of bricks and mortar and other materials needed in the construction process, and the expense of providing infrastructure such as road access, water and power supply, and waste disposal facilities. Indirect costs include the cost of borrowing money to finance the construction and for paying fees or taxes. There are also development costs including those of acquiring the land and the provision of additional infrastructure; in Indonesia, for example, for every one luxury house, a developer is required to build three middle-class houses and six low-income houses (Pangestu 1996). The difference between the cost price and the market price represents the profit (or loss) for a developer.

A key element in land values is location. Early in the nineteenth century, in his seminal work on locational analysis, the geographer Von Thünen postulated that property prices would decay away from city centres. He suggested that location determines the value of land—the further one is away from a city centre, the lower will be agricultural rents and wages. Within cities, certain sites are more prestigious or are more strategically located than others and can therefore command a higher price in the market. A shop at the junction of two streets is likely to attract more passing customers than one located in the middle section of a street and may therefore generate more trade and hence be a more valuable site. A factory located near a good railway or road distribution network will be more valuable than one with less favourable access.

6.4 Land Valuation and Assessment

It is necessary to estimate the value of any land or property when buying, selling, leasing, or taxing it, or when there is a need to calculate the assets held by an individual or business for the purposes of inheritance, bankruptcy, or collateral. Valuations are also needed for investment management and for insurance. Good valuations guide the market towards fair prices and allow informed decisions to be made about the efficient use of resources. For instance they can assist investors in choosing between property and other investments in the wider market place or guide banks in determining how much they can safely lend to a mortgagee.

Several techniques for estimating the value of a property may be used, for example the comparative method, the profits method, the contractor's method, and the residual method. Each technique serves a different purpose, makes different assumptions, and usually provides a different value.

Market Valuation by Comparisons

The simplest way to determine the market value of a property is to identify another property that has recently been sold and which has identical characteristics. The procedures that need to be followed are:

1. inspect the property;
2. analyse the data;
3. calculate the value.

Thus, if a house has recently sold for $100,000 and an identical house next door comes up for sale, it may also expect to fetch $100,000. In practice there is no such thing as a house with identical characteristics as no two properties share exactly the same location. Adjustments can be made if the second house has more or less floor space on a percentage basis or if the first house was sold some time ago since when property prices in general may have risen by, for example, 5 per cent; in this case the estimated value would be $105,000.

Changes over time can be monitored by recording a large volume of property sales and then using such techniques as multiple regression analysis to quantify the trends. Such techniques have produced good results for instance with agricultural land but modelling techniques for the valuation of urban properties are in general rudimentary and not very reliable.

In many countries, especially those in economic transition, data on comparable sales are not readily available and this makes comparisons difficult. Elsewhere, data may not be released because of confidentiality, to gain competitive advantage or through conservatism while that officially held by the land registration or cadastral authorities may not be a reliable indicator of market prices. In inefficient markets where supply and demand are unknown, pricing may be little more than guesswork; even in well organized markets, estimates of property values often vary by 10 per cent. In the housing market, valuations often produce no more than a figure at which negotiations can begin. On the other hand, as has been pointed out, 'from an economic standpoint the existence of valuers can be justified if the improved workings of the market results in efficiency gains which outweigh the resources and costs employed in valuation' (Estates Gazette 1996).

Land and property taxes need to be assessed on a more objective basis than might appear to be the case in the open market. Where sufficient data are available, mass appraisal techniques can be used to provide a basis for taxation. In the United Kingdom, following the failure of the Poll Tax, which was to have been paid by all individuals who were 18 years old or over, a new property tax, known as the Council Tax, was introduced in 1993. The tax is a local tax, based on the relative value of each domestic property in the local area (there is a separate tax for non-domestic properties such as business premises). The estimated market value of each property was determined on the basis of what it might reasonably be expected to sell for on 1 April 1991 and each property was placed in a band.

TABLE **6.1** Council Tax bands for England

Valuation Band	Range of values
A	£40,000 and below
B	over £40,000 and below £52,000
C	over £52,000 and below £68,000
D	over £68,000 and below £88,000
E	over £88,000 and below £120,000
F	over £120,000 and below £160,000
G	over £160,000 and below £320,000
H	over £320,000

The valuation of all domestic properties was completed at an overall cost of just under £75 million or about £3.50 per property. The work was undertaken mainly by private sector surveyors who assisted the central government Valuation Office Agency. The surveyors had no right to access properties; hence the valuations were derived from existing records and external inspection. Certain matters such as the state of repair of buildings were ignored and properties were valued as if they were in good condition, based on their age, character, and location. Approximately 21 million properties were valued by this means within one year and less than 5 per cent gave rise to appeals (Sanderson 1997).

Valuation on the Basis of Profits

While details of property sales are usually in the public domain, indeed some sales are through public auctions, the renting of property is often a more private affair. Rental values are not necessarily publicly available, yet many governments choose to raise taxes on the basis of the assessed rental value of property, as this is the actual or notional income that a landowner should derive from the property. Landlords will receive rent from their tenants while ordinary house owners could in theory rent out their home and hence may be deemed to be receiving the equivalent rent. This can then be taxed in the form of 'rates'.

Alternatively, governments may tax the estimated yield from the land so that farmers are taxed on the basis of what they ought to be able to produce and sell. The origin of the cadastre in the Indian subcontinent was to support the assessment of potential yields taking into account the types and number of crops per year that could be grown, estimated market prices, and the cost of getting the food to market. The estimated earnings after all expenses had been deducted were then taxed.

More valuable properties will attract higher rents and there is thus a relationship between land and property values and rental values. Thus if the annual rental values can be determined, it should be possible to estimate the capital value of a property. In what is sometimes known as the investment method of valuation, properties purchased to provide an income flow are valued by the purchaser in accordance with the calculated potential profit. If an investor requires a certain level of profit from a property, then given current interest rates and the cost of borrowing money to purchase the property and the income that should be derivable from the property, the maximum purchase price can be established. If the vendor will not sell for this price then the purchaser would not make the profit sought and would not buy. Thus an estimate of the value of the property can be computed, the calculations being supported by valuation tables and increasingly by computer software programmes.

Valuation by Construction Costs

A third method of valuation is to determine the construction costs for a site. Known as the Contractor's Method, it assumes that the cost of the site can be determined from other sources. The approach may be used for insurance purposes or for valuing public buildings such as town halls, hospitals, or police stations. It is based on the assumption that the cost of the site plus the cost of building (bricks, mortar, labour, etc.) less depreciation of the building is equal to the value of the property. Thus an old public building on land now worth $500,000 with 5,000 square metres and current building costs of $1,000 per square metre, but with an estimated depreciation of 40 per cent would be worth $3,500,000. The site, area and current building costs should be known to a reasonable degree of accuracy but the figure for depreciation and obsolescence is largely a matter of informed guesswork. Nevertheless an estimate of value can be derived.

Site Valuation by the Residual Method

This method may be used by a purchaser to establish what a site would be worth, and hence what profit could be gained if it were subject to a proposed development. The technique can also be used to calculate tax levels since, by taxing the land at its hypothetical developed value, the landowner will be encouraged to use the land or property more productively. The value of the completed development is sometimes referred to as the Gross Development Value or Gross Realization.

From a purchaser's point of view, a property currently on the market for $80,000 might, with an investment of $30,000 (for instance in improving the fabric of the building), become worth $130,000. The purchaser could potentially make a profit of $20,000. Thus the property may be perceived by the purchaser as being worth more than its present market price; put another way, the current market price is only part of the purchaser's consideration in deciding whether to buy.

From a local authority's perspective, a vacant and derelict site may be damaging the local environment. The site in its present state may be worth only $20,000 but if fully developed in accordance with the value of neighbouring property, might be worth $150,000. By taxing the owner of the site on the basis that it is worth $150,000 the owner will be encouraged to sell or to develop the land. Thus the tax becomes an instrument of planning control.

6.5 Creating a Fiscal Cadastre

Of the four categories of cadastre (juridical, fiscal, land-use, and multipurpose), the fiscal cadastre is most directly concerned with land markets. It is an instrument for administering both tax policy and land policy. The initial creation of a fiscal cadastre may be expensive since a primary requirement is a set of current property maps that provide an index for compiling and maintaining valuation information. The maps are necessary to ensure that all parcels are identified and that no parcel is taxed more than once. The approximate size, shape, and location of the parcel, as depicted on the map, may be used in the actual valuation process.

The creation of a fiscal cadastre involves the:

1. identification and mapping of all properties which are to be subject to tax;
2. classification of each property in accordance with an agreed set of

characteristics relating to such matters as its use, size, type of construction, and improvements;
3. collection and analysis of relevant market data. This may include data on sales prices, rental charges or building maintenance costs, together with details of the dates when these applied;
4. determination of the value of each parcel and its associated buildings;
5. identification of the person or persons who will be responsible for paying the tax; it is also important to identify the legal owner in case the property is forfeited through non-payment of tax although the seizure of property is generally too drastic a measure for most governments to take unless there are very exceptional circumstances;
6. preparation of the valuation roll;
7. notification of the individual property taxpayer of what has to be paid and collection of the appropriate taxes; and
8. provision of procedures so that any citizen who believes that the assessment is incorrect can appeal for it to be reviewed.

Most countries have some rudimentary form of land register or land use cadastre; many however lack a fiscal cadastre. Where rudimentary land taxes are already accepted, the expense of compiling more comprehensive records can be more than offset by the revenue that results from improved tax collection. In Indonesia, for example, a pilot study on developing a land information system resulted in a 50 per cent increase in the tax gathered because many names had been omitted from the existing textually based registers.

Where the recording of land rights in the land registers or land books and the operations of the cadastre are already carried out efficiently, the integration of the two sets of records can bring tangible benefits. In Austria for example the cadastre has been separate from the land registers; the two systems have now been linked through computerization, resulting in savings through less duplication and the provision of a more efficient service. However, such an approach can raise major institutional and political issues, especially where there is confidentiality of the data held for fiscal purposes. Difficulties can also arise where the ownership parcel differs from that used for tax purposes so that the juridical cadastre fails to match the fiscal. None the less, if such problems can be overcome, the potential for saving in administrative costs and for generating marketable information is considerable.

References

de Almeida, O. T., and Uhl, C. (1995), 'Brazil's Rural Land Tax', *Land Use Policy*, 12/2.

Estates Gazette (1996), 'The Economic Role of the Valuer', *Estates Gazette*, 9633.

George, H. (1879), *Progress and Poverty* (San Francisco: W. L. Hilton).

Netzer, R. (1998), 'Land Value Taxation: Could it Work Today?', *Land Lines: Lincoln Institute of Land Policy*, 10/2.

Organization for Economic Co-operation and Development (OECD) (1992), *Urban Land Markets: Policies for the 1990s* (Paris: OECD).

Pangestu, P. (1996), 'New Town Development and the Role of the Private Sector in Indonesia: Land and Rural/Urban Linkages in the Twenty-First Century', *Proceedings of Dialogue 6* held in collaboration with the Habitat II Conference, Istanbul.

Pieri, C., Dumanski, J., Hamblin, A., and Young, A. (1995), *Land Quality Indicators*, World Bank Discussion paper No. 315 (Washington: World Bank).

Roakes, S. (1996), 'Reconsidering Land Value Taxation', *Land Use Policy*, 13/4.

Royal Institute of Chartered Surveyors (RICS) (1996), *Appraisal and Valuation Manual* (Red Book) (London: RICS).

Sanderson, P. (1997), 'Using the Private Sector for Property Tax Development', in *Private and Public Sector Co-operation in National Land Tenure Development in Central and Eastern Europe* (Rome: FAO) (papers of conference in Berfino, Italy, available on Seminar's Web site http://www.fao.org).

United Nations Economic Commission for Europe (UNECE) (1996), *Land Administration Guidelines: With Special Reference to Countries in Transition* (New York and Geneva: United Nations).

7

Land Use and Land Use Control

7.1 Land Use and Sustainability

One fundamental objective of good land administration is to ensure sustainable development. All landscapes change over time, whether through human interference or by natural processes. It is essential that these changes are monitored and understood and that the uses to which the land is put are sustainable and that development: 'meets the needs of the present without compromising the ability of future generations to meet their own needs' (WCED 1987). To be sustainable, development must meet both economic objectives related to returns on investment and social objectives related to meeting people's material, cultural, and spiritual needs. Development must also be ecologically sustainable such that it preserves the long-term viability of the underlying ecosystem (Dalal-Clayton et al. 1994).

In 1992, virtually all the nations of the world signed up to the United Nations Agenda for the twenty-first century, known as Agenda 21 (UNCED 1992) and committed themselves to develop national strategies for the good management of the environment. Since then there has been a growing awareness of the need for sustainability.

From a land administration perspective, the challenge for sustainable development will require in part reconciling perceptions of the nature and value of land which range from:

1. a commodity that can be bought and sold;
2. a functional space;
3. an aesthetic resource; to
4. the setting for activities, such as 'the patterned ways in which households, firms and institutions pursue their daily affairs' (Kaiser et al. 1995).

Of special importance are the ways in which the land is or will be used. The term 'land use' has many different interpretations but in the present context it may be defined as the economic and cultural activities practised upon the land. As such, it is distinguished from land cover, which denotes the physical state of the land and describes the quantity and type of vegetation and other material that occurs on the earth's surface.

Land use and land cover are interconnected by human actions that directly alter the physical environment such as bio-mass burning, irrigation, reforestation, and the applications of fertilizers. These represent the intersection between physical processes and human behaviour:

> On one side, these proximate sources produce land cover changes, or alterations of the properties of the land surface, taking either the form of a conversion or modification, and they may lead to secondary environmental impacts (such as soil erosion, microclimatic change, trace gas emissions, changes in water quality, etc.); on the other side, the proximate sources reflect human goals mirrored in land uses and land use changes (Meyer and Turner 1994).

Land use and land cover are generally described in terms of hierarchical classification schemes that provide progressively more detailed levels of description. For example, the US Geological Survey has developed a series of levels for its integrated land use and land cover classification system that encompasses:

- Level I
  —general categories such as urban, forest, agriculture, and wetland;
- Level II
  —more specific categories such as types of forest land (deciduous, evergreen, and mixed); and
- Level III
  —even more detailed descriptions such as specific tree species (Anderson et al. 1976).

Some classification systems focus on the physical characteristics of the land while others are more highly oriented towards human activities; some concentrate on rural areas and activities while some are more concerned with a detailed classification of urban land use to support the administration of town planning legislation. A lack of a standard classification system can mean, as has happened in the United Kingdom, that two adjoining local authorities use the same word to describe different activities or different words for the same activity, often leading to considerable confusion.

7.2 Land Use Planning in a Historical Context

Efforts to control the way in which land is used have a long history. Some of the earliest recorded land surveys were for the control of land use in Mesopotamia and the Nile delta. Many systems for controlling customary land use have origins that date back millennia, such as those practised in the Indian subcontinent where an orderly pattern for the control of how and by whom the land may be used dates back thousands of years. The open field system established by the Saxon settlers in England, for example, 'reflected a rigidly controlled pattern of land use: arable fields, divided into furlongs, subdivided holdings, all cultivated in accordance with a traditional rotation of crops and all subject to common grazing at certain seasons of the year, with similar rights in the meadows and common rights in the waste' (Hart 1964).

Modern concepts of land use planning and control date back to the middle of the nineteenth century and to the rapid growth of urban populations resulting from industrialization in Europe. Increasing concern about public health, fire safety, and transportation led to local authorities being delegated responsibility for drainage, water supply, and roads and more generally, in the language of the day, for promoting the health, safety, and general welfare of their communities. The first major steps were taken in the 1850s when London and Paris undertook ambitious urban renewal programmes in their city centres. At about the same time, local governments in Germany were given the power to establish street alignments, regulate building construction, and control land use. Land use planning in North America dates from the 1860s, while in 1875 New Zealand enacted 'Plans of Towns' legislation that set out requirements for approving town plans, regulating the width of streets, and requiring that space be reserved for recreational facilities, gravel pits, rubbish tips, and other amenities.

These developments were underpinned by the ideas of pioneer planners such as Ebenezer Howard, who was the British originator of the garden city, Thomas Adams, a Canadian surveyor and landscape architect, and Patrick Geddes, who 'appreciated that human life and its individual and collective environments, both natural and manufactured, are all intertwined, and that improving cities means understanding them and planning for that reality rather than substituting another urban form' (Hodge 1986).

Geddes has been credited with saying 'no plan before survey', stressing the importance of undertaking surveys of a town's geography, economy, and social conditions and presenting them in the form of a 'civic exhibition'. In his view this would engage both officials and citizens in appreciating and resolving the problems of the community.

By the turn of the century, city planning was becoming a recognized discipline, with Germany providing much of its intellectual and technical foundation. During the first two decades of the twentieth century much of the core legislation that underpins Anglo-American land use control regimes was first introduced. Before this time, regulation was founded in such common law principles as the concept of nuisance which held that a property owner could not use his property in such a manner as to injure that of a contiguous owner. These were limited and mostly 'negative' regulatory powers that were increasingly seen as ineffective in dealing with the health and social problems of the day.

During this period, zoning was probably the most important of the new land use control instruments to be introduced. 'Zoning lent modern scientific legitimacy to the law's traditional concern with conflict avoidance. The method was simple and appealingly balanced, altogether objective and efficient in appearance' (Plotkin 1987). Zoning, based on the premiss of being able to design and regulate the most compatible patterns of land use in a city, was 'simply the geographical extension of the law of nuisance from a few adjacent properties to cover neighbourhoods and districts' (Hodge 1986). Zoning by-laws were first developed in Germany towards the end of the nineteenth century in an effort to improve housing conditions. They were almost immediately, although unsuccessfully, challenged in the courts on the grounds that they diminished the rights of property owners. In the United States, the first comprehensive zoning ordinance was adopted by New York City in 1916 and a decade later the US Supreme Court (1926) subsequently upheld the constitutionality of zoning in the famous case of the *Village of Euclid, Ohio* v. *Ambler Realty Co*.

After the Second World War, support for community planning and land use control grew rapidly in the developed world. Governments increasingly had to contend with the post-war economic boom and urbanization. They also had to provide for the capital investment in infrastructure and community facilities that had been postponed during the depression and war years (or, in the case of Europe, to redevelop the city centres that had been bombed or were obsolete). At the heart of the post-war planning process was the land use plan, an official document providing a generalized design for the future use of land, covering both private land use and public facilities. The format of the land use plan typically included 'a statement of objectives, a description of existing conditions and future needs for space and services, and finally the mapped proposal for the future development of the community, together with a program for implementing the plan (customarily including zoning, subdivision control, a housing code, a public works expenditure program, an urban renewal program, and other regulations and development measures)' (Kaiser and Godschalk 1995).

Kaiser and Godschalk (1995) have identified four major land use planning and control strategies that have gained widespread currency over the past two decades (see Table 7.1):

1. the land use design, which entails a detailed mapping of future land use arrangements and which is similar to traditional land use planning but with the inclusion of environmental processes and linkages to a community's fiscal capacity;
2. the land classification plan, which provides a more general map of potential growth areas rather than a detailed land use analysis;

TABLE 7.1 Land use planning and control strategies

Feature of plans	1950s general plan	Contemporary prototype plans				1990s Hybrid design-policy-management plan
		Land use design	Land classification plan	Verbal policy plan	Development management plan	
Land use maps	Detailed	Detailed	General	No	By growth areas	General and area specific
Nature of recommendations	General community goals	Land use policies & objectives	Growth locations & incentives	Variety of community policies	Specific management actions	Policy and actions
Time horizon	Long range	Long range	Long range	Intermediate range	Short range	Short and long range
Link to implementation	Very weak	Weak to moderate	Moderate	Moderate	Strong	Moderate to strong
Public participation	Pro-forma	Active	Moderate	Moderate	Active	Active
Capital improvements	Advisory	Recommended	Recommended	Recommended	Required	Recommended to required
Land use/transportation linkage	Moderate	Strong	Weak	Varies	Strong	Strong
Environmental protection	Weak	Moderate	Strong	Varies	Varies	Strong
Social policy linkage	Weak	Weak	Weak	Moderate to strong	Weak	Moderate

Source: Kaiser and Godschalk 1995.

3. the verbal policy plan, which focuses on written statements of goals and policies rather than mapping specific land use patterns or implementation strategies; and
4. the development management plan, which provides a specific action programme for guiding development (such as public investment strategies, development codes, programmes for extending infrastructure and services, etc.).

Regardless of the strategy chosen, the challenge for politicians and planning officials has increasingly been to balance the economic, social, and environmental demands being placed upon the land and to involve individual citizens and the community as a whole in the planning process.

7.3 Land Use Control

Governments determine how land is to be developed and used in a variety of ways, including direct acquisition, the provision of incentives, and through regulations. The acquisition of lands for public purposes has traditionally been accomplished through purchase or expropriation, and then managed in a form not dissimilar to that of a private landowner. Lands may also be acquired in trust either by a public agency or by a private non-profit corporation and preserved for some biological, historical, or aesthetic purpose. A limited form of land trust is the conservation easement, a non-possessory interest in the land that imposes limits or affirmative obligations on how the land may be used (Cheever 1996).

Various incentive instruments are available for encouraging land to be used or developed in support of public policy objectives. Many jurisdictions, for example, have developed incentive strategies for preserving prime agricultural lands. The goals of these strategies have included limiting the conversion of agricultural lands to other uses, minimizing the forms of land use that are incompatible with agricultural operations, and promoting the long-term sustainability of the agricultural community. Incentive measures include the creation of transfer development credits (allowing farmers to be compensated in the market place for forgone development rights), the development of preferential taxation schemes (for example, by assessing farmland for tax purposes on the basis of use value rather than market value), and the establishment of various agricultural support programmes (such as transportation and irrigation subsidies).

There are two basic approaches to regulating how land is developed and used: by way of legislation applying to all properties uniformly, or by way of a permit system in which a property owner must make application at the time of a proposed development. The most common forms of land use control are:

- zoning;
- site plan control;
- building regulations;
- development control.

Zoning entails the delineation of a community into districts or zones within which certain activities are permitted and others are prohibited. It provides a set of standards with respect to the specific use or uses to which a parcel of land may be put, and the size, type, and placement of improvements on that parcel.

These standards are generally made explicit in the text of the zoning regulations (such as the municipal by-laws) and an accompanying map specifies the boundaries of each zone. Zoning legislation may also regulate the density of population in a given district by limiting the number of dwelling units or occupant density. Special districts may also be identified that contain distinctive local features for historical, environmental, functional, or aesthetic reasons.

Within the urban context, three basic land use categories are usually identified: residential, commercial, or industrial. For each of these categories, there will often be a series of zones depending on the specific nature of the activities, the density of development, and the parcel size.

Site plan controls, including subdivision regulations, provide rules for determining the way that land is to be subdivided and how it is to be prepared. The controls govern building, constructing public improvements such as streets, pavements, sewers, and lights, and setting aside land for public purposes such as schools and parks. Site plan or subdivision control has both a substantive and a procedural component. Substantively, it promotes the development of a quality physical environment according to recognized planning and engineering standards. Procedurally, it provides a monitoring process, with prescribed and regulated steps, for all those public agencies having an interest in how the land is to be developed. Figure 7.1 describes schematically the typical steps involved, and the kinds of surveys required, in preparing a land subdivision.

There are several common procedures to most site or subdivision control (Hodge 1986), including:

1. A specified approval process. This provides details of how a developer must apply to have land subdivided, the steps that will be taken to review the subdivision plan and often the maximum amount of time the process can take.
2. Plan circulation. Once a draft site plan has been prepared (usually by a surveyor or engineer), it will typically be circulated to a number of public agencies for review (including those responsible for transportation, health, housing, and utilities).

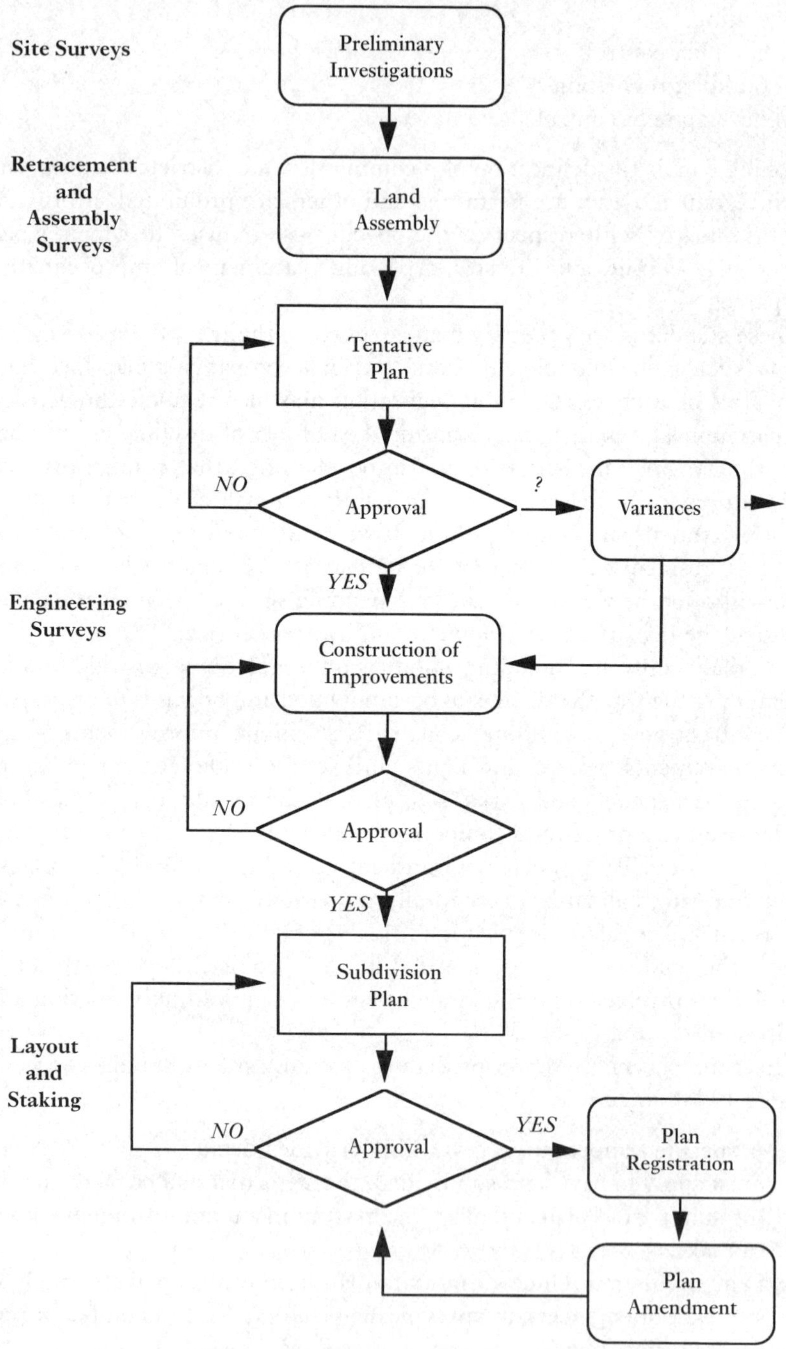

FIG. 7.1 Simplified subdivision development process

3. Conditions for approval. In addition to meeting specified planning and engineering standards, the developer will also generally be required to set aside lands for roads, parks, utilities, and other public purposes. Also in order to ensure the ongoing provision of services, the developer is often required to enter into a formal contract with the local government.
4. Final plan approval. The site or subdivision plan is not effective until it has received final planning approval and been registered. Upon registration the new parcels are officially created and the agreements regarding services and standards become binding.

In addition to overall planning controls, building controls are employed to govern the way in which buildings are constructed and maintained. These date back many centuries—London, for example, has had building controls since the twelfth century, with substantial regulations being adopted after the Great Fire of 1666. Building controls are designed to ensure a safe and healthy environment by regulating the design, construction, erection, placement, occupancy, and change of occupancy classification of existing buildings and the work necessary to correct unsafe conditions in existing buildings (RSNS 1989). Figure 7.2 provides an example of a typical building regulatory regime.

Controls are typically administered through a building permit process, with building control surveyors making site inspections to ensure compliance with the regulations. The effective enforcement of building regulations can be a complex and expensive process, often resulting in the regulations being administered only in relatively wealthy urban settings. Recent efforts to simplify the process (and extend their reach) have concentrated on redefining and limiting the enforcement function either through making on-site inspections to audit what is going on or by relying primarily on the certification and assurance of the private sector building industry (Ministry of Municipal Affairs 1994).

Development control, or development permit, has a strong British tradition. British planning practice:

> never embraced zoning, preferring to leave responsibility for plan implementation in the hands of local administrators. Development control was considered particularly useful for the time period when a formal community plan was being prepared, especially in fast-growing cities. Each proposal for new development, buildings or subdivisions, could be reviewed to ensure consistency with the aims of the emerging plan, and a development permit issued (Hodge 1986).

In many countries, development control focuses primarily on the construction or alteration of buildings and regulates the placement of the building on the site (the so-called building envelope), its appearance, and its relation to surrounding buildings and streets.

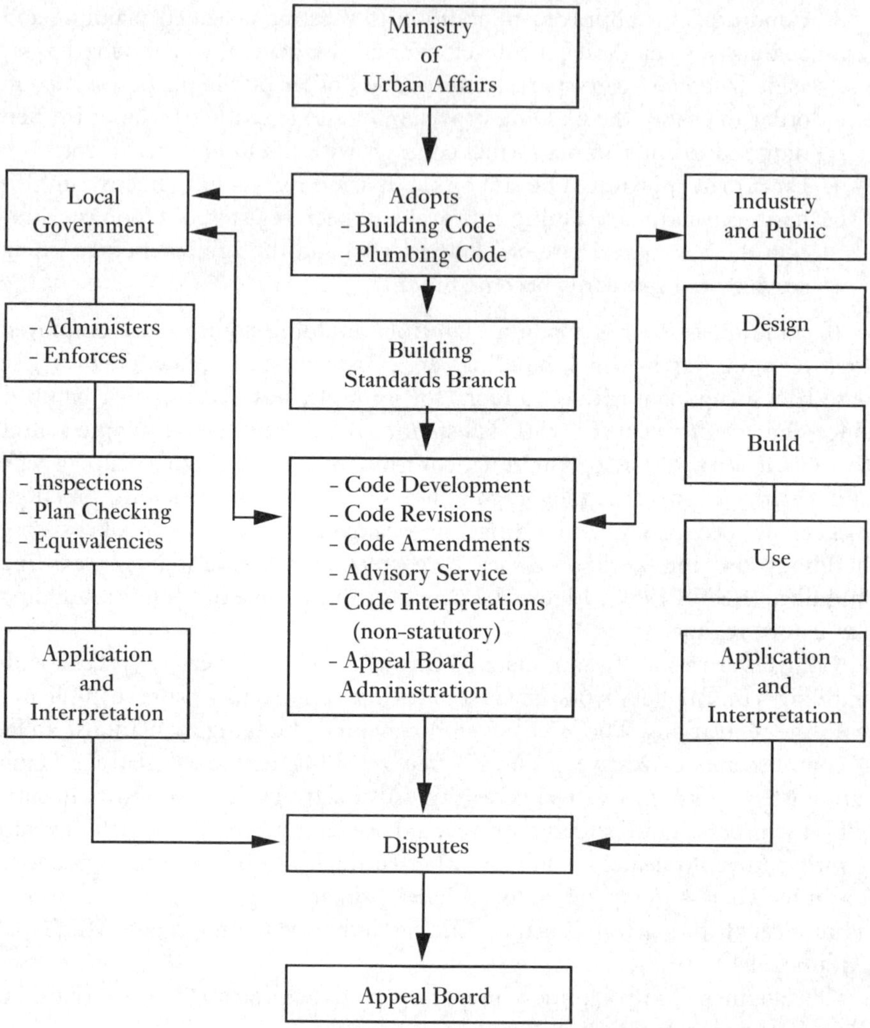

FIG. 7.2 Building regulatory regime

7.4 Monitoring Environmental Impact

During the 1970s, as people became more concerned about the environmental damage to which the world was being subjected, techniques such as Environmental Impact Assessment (EIA) became fashionable. Prior to this time, development was assessed mainly on the basis of engineering and economic feasibility (often through the use of cost–benefit analysis) with limited concern for the impact on the environment at large. Early attempts at EIA were more concerned with the impact that projects might have on the ecology of areas immediately affected, especially on local flora and fauna such as birds and fish. By the mid-1980s the whole basis of sustainability was under review and wider concerns for matters such as global warming were being expressed. Concern peaked in 1992 with the Rio Conference on the Environment and the adoption of Agenda 21, since when most governments have been seeking ways to respond to the requirements of that Agenda at minimum cost.

Current approaches seek to balance environmental analysis with the objectives of resource management. Because a full analysis is often a very expensive and time-consuming process, most projects are subjected to an initial screening process to determine whether they are likely to create any significant environmental problems; many, from past experience, do not and planning approval can be given without further concern. Those projects that may seriously impact on the environment can then be subjected to further scrutiny and an initial impact assessment. After this, the relevant authority charged with approving the project may decide that, in the light of the initial study, the balance of risk favours the developer and the proposal can go forward. Only where uncertainty or concern remains will a full impact assessment generally be called for.

In presenting an environmental impact assessment it is usually necessary to:

- document the background to the project and the reasons why it is taking place;
- review the basic legal and development policy framework within which the project will be undertaken;
- identify the possible conflicts that may arise, who will be affected and over what time scale;
- examine options available and alternative ways of achieving the project's goals;
- determine the data that will need to be collected both about the project and its impact on the environment;
- analyse the potential impact identifying both positive and negative features and assessing the probability associated with each risk or benefit factor;
- establish what monitoring procedures are necessary to ensure that the project progresses as intended; and
- draw conclusions and make recommendations on all relevant technical, political, and economic issues.

From a land administration perspective, the information infrastructure should provide details of the use rights and legal restrictions that may apply to the land. It should also indicate the current land use and provide data on the changing patterns of land use in the area, thus supporting the monitoring process. While most cadastral systems are currently unable to provide such data, as they have not been kept up to date and the data are not yet sufficiently accessible, computerization and re-engineering (see Chapter 9) are creating new opportunities to overcome these problems.

7.5 Some Regional Perspectives

While much of the post-war focus of land use planning and control has related to urban areas, increasing attention is also being given to both the rural environment and to the broader regional context for planning and controlling land use. Already, in the nineteenth century, concern about increasing threats to the natural environment as a consequence of rapid industrialization led planners such as Frederick LePlay in France and later Patrick Geddes in Britain to promote ways for achieving a balance between human activity and nature. Geddes identified three factors to be taken into account in physical planning: Folk (the people of the region), Work (the economy of the region), and Place (the geographical dimensions of the region) (Glikson 1955). Out of this thinking emerged the concept of regional planning as a means for addressing two sets of concerns: the growth and spread of cities into the countryside; and preserving and enhancing rural economies.

The Garden City movement, promoted in Britain by Ebenezer Howard in the 1890s, exemplified an early initiative to address the problems of urban sprawl. The Tennessee Valley Authority (TVA) provided a famous example from the 1930s of planning in support of rural economic and social development. The TVA's mandate in the depths of the depression was to rehabilitate a poor, rural region encompassing several thousand square kilometres in the south-eastern United States through the construction of dams and power facilities, reforestation, improvements to agriculture, and the building of new towns.

During the past three decades regional planning has seen the introduction of three overlapping approaches: (i) watershed planning; (ii) rural resources planning; and (iii) rural economic development (Hodge 1986). Watershed planning, in the tradition of the TVA, has addressed the issues of developing, controlling, and using water resources to generate electricity, prevent floods, and to provide irrigation. Rural resources planning has been concerned with achieving a balance between urban and rural land needs and with the preservation of rural life while rural economic development has focused on advancing strategies for economic sustainability and minimizing regional economic disparities.

While substantively different, each of these approaches has been concerned with planning for large geographical areas, and with locating activities and developments (as opposed to allocating space amongst competing land uses in urban areas). Regional planning has also largely been undertaken in an institutional setting of limited government control, with much of the planning function being of an advisory nature only.

During the last decade, there has been a renewed interest in promoting comprehensive and integrated approaches to regional planning, especially as it relates to the concept of integrating the economic, social, and environmental dimensions of development. Much of this interest can be traced to the publication of the report of the World Commission on Environment and Development (1987). New Zealand, for example, began a process of law reform in 1988 in support of sustainable management that sought to manage the 'use, development and protection of natural and physical resources in a way, or at a rate, which enables people to meet their needs without unduly compromising the ability of future generations to meet their own needs' (Bush-King 1991).

The small, rural Canadian province of New Brunswick provides a more recent example, with its emphasis on integrated rural planning and control. In 1992 the provincial government established a Commission on Land Use and the Rural Environment with a mandate to explore the various issues facing land use and the rural environment in New Brunswick. The challenges facing the province included:

- the absence of a comprehensive institutional framework for rural development and resource management;
- the absence of mechanisms for developing comprehensive and integrated plans as well as procedures for conflict resolution;
- a traditional reliance on urban planning methods to deal with rural issues;
- a growing need to address uncontrolled settlement in the form of sprawl and ribbon development;
- the need to embrace the concept of sustainable development; and
- a requirement to more effectively involve the citizens of the province in the decision-making process.

After extensive research and public consultation, the Commission recommended a sweeping series of policies and strategies to protect and enhance the quality of the rural environment while fostering sustainable development and responsible use of the Province's natural resources (Stelling 1997). Beyond this, the province has pioneered the development of new, 'one-stop' facilities for the delivery of public services and the use of Internet technology for providing public access to land use planning information (Finley et al. 1998).

7.6 Public Participation

Since the 1970s there has been an increasing emphasis on engaging the public in the land use decision-making process. This has occurred in large measure due to an increased criticism of the remoteness of big government, a growing awareness of the importance of non-economic values in making decisions, and a heightened demand for more participatory democracy. In response, the role of planners and administrators is gradually changing from that of arbiter of the public interest to one of facilitating amongst the range of interest groups that collectively define the public interest. Increasingly it is recognized that decisions that 'fail to balance public interests through a lack of reliable information or as a result of relevant interests not being given sufficient attention lead to instability, continuing conflict, lack of sustainability, and long-term inefficiencies in the use of government personnel and funding' (CORE 1995).

The Canadian province of British Columbia has recently developed an innovative strategy for effecting meaningful public participation in its provincial land use strategy. This has included defining the appropriate public rights and responsibilities (with the right of participation balanced by the responsibility to participate in good faith); preparing a policy for promoting meaningful participation; developing strategies for consensus-seeking, multi-party public negotiations; creating a framework for aboriginal participation; and providing for the establishment of community resource boards. Community resource boards, or round tables, represent a wide range of community interests and seek to provide a means of working out cooperative solutions on a variety of important land use issues. They are 'voluntary and purpose-driven, open to participation of all interests and consensus seeking. Through inclusion and cooperation, community resource boards harness democratic energy at the local level, and provide government with an efficient forum for advice on decisions that require public information, balance and support' (CORE 1995).

7.7 Managing Land Use in Developing Countries

The developing world has had to face immense land use planning and regulatory challenges in response to rapid population growth, industrialization, and urbanization. Governments have largely failed to meet these challenges, in part because land use planning and regulations have been rigid and cumbersome, imposing high costs on the builder or developer, and rarely being effectively enforced.

A World Bank study identified several specific areas of concern:

1. most regulations are based on outdated and inappropriate planning legislation, with a heavy emphasis on centralized control;
2. master plans take too long to prepare and rarely address the financial implications associated with their implementation;

3. land use planning and administration functions are often institutionally fragmented across a number of ministries; also, these functions have traditionally been isolated from their economic counterparts;
4. control over development is enforced primarily by extensive bureaucratic approval procedures. In many countries, such as Ghana, Pakistan, and Peru, the approval process can take anywhere from two to seven years (Farvacque and McAuslan 1992).

The consequence has been that in

> the majority of cities with master plans, the supply of shelter for the low-income population is built in spite of the master plan, not because of it. In brief, master planning and comprehensive planning techniques are primarily concerned with the product rather than process and do not adequately address implementation issues, the increasing complexity of land markets, the role of the public sector versus private sector actions, and the links between spatial and financial planning (Farvacque and McAuslan 1992).

Squatter settlements have posed special challenges. Typically, the housing in urban slums is made from flimsy materials such as cardboard, plastic sheeting, and scrap metal. They are often built in hazardous areas where there are a multitude of dangers such as the potential for landslides (as happened in the 1998 devastation in Central America due to Hurricane Mitch) and there is little access to clean water and waste disposal facilities. The area covered by such housing is usually at very high density and expanding fast. In Asia, squatter settlements already provide accommodation for a quarter of the urban population.

Without formal recognition, these settlements rarely benefit from municipal services such as electricity, water supply, and rubbish collection. Often the most important first step in providing for decent shelter and healthy living conditions is speeding up the process of legal recognition.

> If slum dwellers know they may be evicted at any time, they are unlikely to invest much in their homes. Conversely, slum projects in Latin America and Asia suggest that granting secure tenure prompts locals to invest in safer homes and better sanitation. That does not mean squatters should automatically be given ownership of land they occupy; but it does raise hopes that, once a government has settled the issue of land title, some of the worst environmental problems may disappear (Anon. 1998).

The World Bank has identified seven basic goals for improving urban land use planning in developing countries:

1. better coordination in land management;
2. integration of spatial planning with financial, sectoral, and institutional planning;
3. definition of a collective rationale for intervening in the land markets;

4. better protection of the environment;
5. economic efficiency to maximize the benefits of urban development;
6. equity considerations in support of poor households;
7. cost effectiveness to minimize public costs and to recover them (Farvacque and McAuslan 1992).

These goals must be supported with a regulatory framework that is linked to national economic and social objectives and that is capable of effective implementation. In addition: 'adequate information is required to define the framework, to modify it in the light of changing circumstances, and particularly to adjust it to take account of the effects of rapid urban growth' (Courtney 1983).

The increasing severity of economic, social, and environmental problems in the rural areas is also of major concern. Current efforts to address a multitude of interrelated problems, including deforestation, desertification, water pollution, and urban sprawl have been hindered by a piecemeal and uncoordinated approach. There has often been duplication of effort or conflicting sectoral goals—yet another example of why there is a need to take a more holistic approach to the management of land and land resources.

North-east Brazil presents an important cautionary tale. Chronic poverty in the north-east led the Brazilian government to undertake a series of integrated rural development projects in the 1970s and 1980s. Despite undertaking six major adjustment programmes and expending more than $3.2 billion, the rural poor are little better off than they were two decades ago. A World Bank team evaluating this history concluded that these projects involved:

1. bureaucratic procedures and excessive documentation requirements;
2. delays in project approval and fund disbursements, and funding limitations;
3. limited technical assistance in general and lack of staff to assist communities in preparing and implementing projects;
4. lack of municipal participation and funding;
5. weak community participation, particularly in prioritizing projects;
6. political interference;
7. lack of sustainability built into the projects; and
8. insufficient knowledge of the programme by communities (van Zyl et al. 1995).

The World Bank team went on to formulate a series of design principles for future rural development projects, including:

1. greater decentralization of the financial decision-making to local and regional governments as a means of promoting more efficient programme administration;

2. enhanced participation in the financing of projects to generate a sense of shared ownership and responsibility;
3. clear target setting, especially with respect to reducing poverty, using mechanisms that are simple, explicit, and able to be monitored (van Zyl et al. 1995).

The team also recommended enhancing project sustainability through shared funding arrangements, standardization of project documents and technical designs to reduce unit costs, using technical assistance to augment local capabilities while retaining local commitment and involvement, and developing clearly defined and well-disseminated systems of checks and balances in order to discourage the misappropriation of project funds.

7.8 Global Land Use Challenges

There are two fundamental human sources of global environmental change: industrial metabolism and land use change. Recent modelling of global climate change by the US National Centre for Atmospheric Research in Boulder, Colorado suggests that changes in land use may contribute up to 0.4 degrees in average global temperature change, out of the total average global temperature rise of 0.5 to 0.6 degree over the last century (Bonan 1997).

Despite the importance of land cover and land use to understanding global environmental change, there is still very little useful data:

> There is a lack of common definitions and standards, there is little in the way of systematic and co-ordinated data collection activities (although there are some exceptions such as the UNESCO soil map of the world at scale 1:250,000), there are numerous problems in interpreting that data that are available and there has been virtually no work done on addressing the issue of cumulative change—change that is manifested through the geographically dispersed, progressive result of human activity, such as soil loss, habitat loss, and loss of biological diversity (Robinson et al. 1994).

These challenges are being addressed on two closely related fronts: through the development of global land use/land cover models (GLM); and through the design and funding of new strategies for collecting and managing the data necessary for these models. One example of a GLM currently under development is that proposed by Rayner et al. (1994). This model:

1. incorporates a process for organizing the available data on land use and land cover (and the forces driving change) into a global framework;
2. assesses data adequacy and defines research priorities;
3. provides techniques (that are subject to continuous improvement) for analysing the mechanisms causing land use change and the cumulative impact of global change;

4. produces projections that can be verified or falsified; and
5. examines the effects of regulating land use.

The model requires data on deforestation and regrowth, the expansion and contraction of cropland and pastures, the growth of settlement areas, land degradation, and changes to wetlands.

A number of recent global change initiatives have included components for improving the extent and quality of data collection (primarily through the use of satellite remote sensing). They have also focused on organizing and managing the resulting data sets (typically using grid scales on the order of 200×200 km) and communicating the results to the international community. Examples of such initiatives include:

- US Global Change Research Program;
- World Climate Research Program;
- International Geosphere-Biosphere Program;
- Human Dimensions of Global Environmental Change Program;
- Committee on Earth Observation Satellites;
- Global Mapping Initiative.

More recently, work has begun on exploring the feasibility of integrating these various programmes into a global geo-spatial data infrastructure. The initial emphasis has been on examining the policies, technologies, standards, and human resources which would be required for the effective collection, management, access, delivery, and utilization of geo-spatial data in a global community (Coleman and McLaughlin 1997).

References

Anderson, J. R., et al. (1976), 'A Land Use and Land Cover Classification System for Use with Remote Sensor Data', Washington: US Geological Survey Professional Paper 964.

Anon. (1998), 'Living Dangerously', *The Economist*, 21 Mar.

Bonan, G. (1997), see http://neit.cgd.ucar.edu/index.html

Bush-King, D. (1991), 'The Resource Management Bill: A Brave New World?', *New Zealand Surveyor*, 33/278.

Cheever, F. (1996), 'Public Good and Private Magic in the Law of Land Trusts and Conservation Easements', *Denver University Law Journal*, 73.

Coleman, D., and McLaughlin, J. (1997), 'Defining Global Geospatial Data Infrastructure Components, Stakeholders and Interfaces', *Geomatica*, 52/2.

Commission on Resources and Environment (CORE) (1995), *The Provincial Land Use Strategy: Public Participation*, vol. iii (Victoria, BC: Government of British Columbia).

Courtney, J. (1983), 'Intervention through Land Use Regulation', in Harold Dunkerley (ed.), *Urban Land Policy: Issues and Opportunities* (Oxford: Oxford University Press).

Dalal-Clayton, D. B., Bass, S., Sadler, B., Thomson, K., Sandbrook, R., Robins, N., and Hughes, R. (1994), 'National Sustainable Development Strategies: Experiences and Dilemmas', *Environmental Planning Issues*, 6 (London: IEED).

Farvacque, C., and McAuslan, P. (1992), *Reforming Urban Land Policies and Institutions in Developing Countries* (Washington: World Bank).

Finley, D., Quek, S., McLaughlin, J., and Coleman, D. (1998), 'The Provincial Land Information Infrastructure for New Brunswick: From Early Visions to Design to Reality', *Geomatica*, 52/2.

Glikson, A. (1955), *Regional Planning and Development* (Leiden: A. W. Sijthoff).

Hart, W. O. (1964), 'Control of the Use of Land in English Law', in Charles Haar (ed.), *Law and Land: Anglo-American Planning Practice* (Cambridge, Mass.: Harvard University Press).

Hodge, G. (1986), *Planning Canadian Communities* (Toronto: Methuen).

Kaiser, E., and Godschalk, D. (1995), 'Twentieth Century Land Use Planning'. *Journal of the American Planning Association*, 61/3.

——— and Chapin, S. (1995), *Urban Land Use Planning*, 4th edition (Urbana: University of Illinois Press).

Meyer, W., and Turner, B. L. (eds.) (1994), *Changes in Land Use and Land Cover: A Global Perspective* (Cambridge: Cambridge University Press).

Ministry of Municipal Affairs, British Columbia (1994), *The Building Regulatory System in British Columbia: Options for Renewal* (Victoria, BC: Queen's Printer for British Columbia).

Plotkin, S. (1987), *Keep Out: The Struggle for Land Use Control* (Berkeley and Los Angeles: University of California Press).

Rayner, S., et al. (1994), 'A Wiring Diagram for the Study of Land-Use/Cover Change', in Meyer and Turner 1994.

Revised Statutes of Nova Scotia (RSNS) (1989), Building Code Act, chapter 46.

Robinson, J., et al. (1994), 'Land-Use Cover and Land-Cover Projections', in Meyer and Turner 1994.

Stelling, H. (ed.) (1997), 'Planning Reform in New Brunswick', Special issue of *Plan Canada*, 37/5.

United Nations Conference on Environment and Development (UNCED) (1992), *Agenda 21* (United Nations Conference on Environment and Development, New York: UN General Assembly).

Village of Euclid, Ohio v. *Ambler Realty Co.* 1926 US Supreme Court.

World Commission on Environment and Development (WCED) (1987), *Our Common Future* (Oxford: Oxford University Press).

van Zyl, J., Barbosa, T., Parker, A., and Sonn, L. (1995), 'Decentralized Rural Development and Enhanced Community Participation: A Case Study from Northeast Brazil', Washington: World Bank Policy Research Working Paper 1498.

8

Land Information Management

8.1 Land Information Systems

A key component of land administration is the management of land and property related data. Such data may be held in manual or digital form although, increasingly, all land related records are being computerized for ease of storage and retrieval. Data are raw collections of facts that, from a land administration perspective, may be gathered and written down as numbers and text, for instance in a surveyor's field book, or collected and stored digitally through the use of 'data loggers' and computers. They may also be held graphically as on maps or aerial photographs. Data become information when processed into a form meaningful to a decision-maker. The usefulness of this information will depend upon the quality of the data and especially on the extent to which they are up to date, accurate, complete, comprehensive, understandable, and accessible. Even then, good data do not necessarily produce good management decisions since other factors may be involved, such as the qualities of the data user; the converse is however true, namely that poor quality data will almost certainly result in bad decision-making.

Land and property related data are increasingly managed within formal land information systems (LIS). As with all information systems (see Figure 8.1), LIS use a combination of human and technical resources, together with a set of organizing procedures, to produce information in support of management activities (Dale and McLaughlin 1988).

Increasingly, the technologies that drive the data processing are components of geographic information systems (GIS). There has been much debate about the nature of GIS, some seeing them as sets of hardware, software, and data while others have seen them as all-embracing institutional arrangements of which the technology is only part. In the following discussion, GIS will be treated as the former and restricted to the acquisition and assembly of spatial data; their processing, storage, and maintenance; and their retrieval, analysis, and dissemination. By analogy with the motor car, GIS are the engines that power the car and data are the fuel; operating a transportation system is, however, a more complex process.

GIS are able to process data, changing them into products (such as maps or

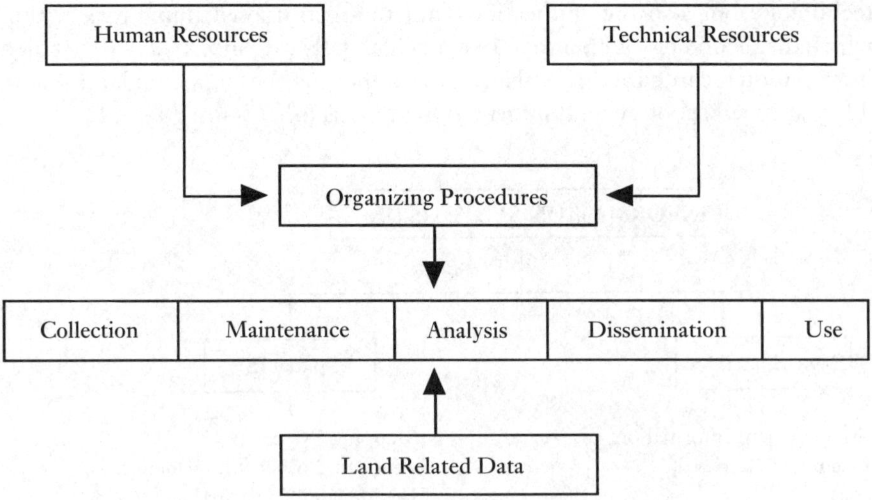

FIG. 8.1 Components of a land information system

tabulations) or services (answering specific queries); they may also be used to analyse the data over time or over space. All objects have attributes that describe their characteristics. In the case of real property these attributes may include a parcel's location, its ownership, official value, and classified use. The data may also include details about relationships to other parcels that are adjacent to or otherwise associated with it. The locational elements of the data may be held in terms of points, lines, or polygons (that is, a series of lines representing areas) and may be displayed:

- geometrically—where the attributes are referenced to a precise spatial framework, such as Cartesian coordinates or grid cells;
- cartographically—where information is graphical, generalized, and symbolized; or
- topologically—where the relative location and the non-metric relationships of various data elements are portrayed, such as which side of a boundary line a feature lies.

GIS allow complex analysis of data to be undertaken, revealing patterns that would not otherwise be detected. To date, such applications of the technology have not had a significant impact on the cadastre because priority has been given to data storage and retrieval rather than to data analysis. Potential applications of GIS abound but only recently have attempts been made to exploit the technology for instance in property valuation (Wyatt 1998).

Whereas GIS has been narrowly defined in terms of computer technology, the term 'land information systems' will be employed to cover not only the technology but also the applications and the institutional framework within which the technology is operated. In particular, LIS are concerned with detailed information recorded at the local level so that they may be mapped at large scales. They are a subset of general information systems (see Figure 8.2).

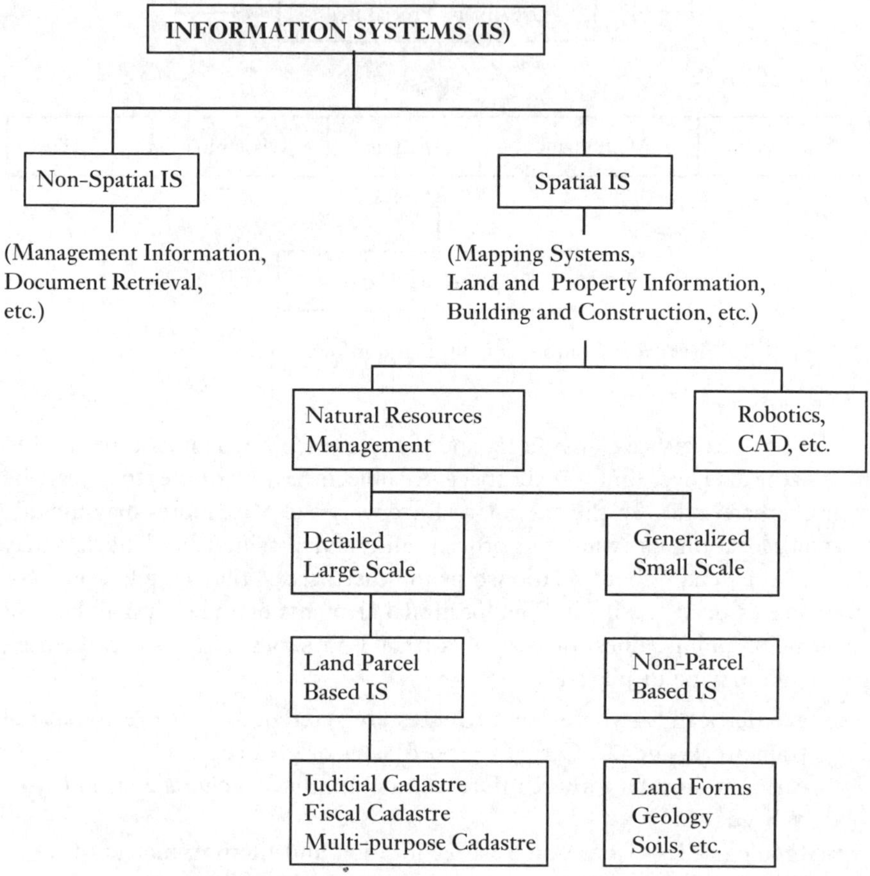

FIG. 8.2 Taxonomy of land information systems

Land information systems may be designed to serve one primary function or they may be multi-functional. Some have been developed to support strategic planning, where the focus is on determining an organization's objectives and on the resources employed to achieve them. Some provide for management control and are concerned with the effective use of resources so as to accomplish the organization's goals. Others have been designed for operational control so that specific tasks can be carried out effectively and efficiently. Each requirement dictates a special set of criteria that determine what information is needed and hence what form the information system should take.

There are at least four different types of land information system, namely:

1. environmental systems, especially those related to soils, geology, water resources, vegetation, and wildlife; these are particularly relevant to rural land management.
2. infrastructure systems, focusing primarily on engineering and utility structures, such as underground services and pipelines; these are particularly important in urban environments.
3. cadastral systems, recording land rights, planning restrictions, and land values; these apply to all land areas especially where land markets operate.
4. socio-economic systems incorporating statistical and census type data.

SPATIAL DATA SETS

Environmental Information	Infrastructure Information	Cadastral Information	Socio-economic Information
Soils, Geology Watercourses Vegetation Wildlife	Utilities Buildings Transport Communications	Tenure Valuation Land Use Law and Order	Health Welfare Population Marketing

FIG. 8.3 Spatial data sets

8.2 Users of Land Information

Within any land information system, there are many possible types of data that may be recorded and many different activities to which they may be applied. From a land administration perspective, the most important systems are those that focus on the cadastral parcel. These include the multi-purpose cadastre which is a special example of a land information system. While some cadastres focus on ownership or value, the multi-purpose cadastre encompasses a more comprehensive set of parcel related data.

Land information systems are required by a wide variety of users, ranging from agencies at all levels of government through existing or prospective landowners to lawyers, surveyors, valuers, real estate managers, and retailers. Examples of uses and users of parcel based data are shown in Table 8.1

TABLE 8.1 Selected users and uses of spatial data

Use of Data	User			
	Central government	Local government	Utilities	Retailers
Asset management	**	**	**	**
Building permits		**	**	
Cultural features	**	**	**	**
Customer identification		**	**	**
Development control	**	**	**	**
Environmental impact	**	**	**	
Health & safety	**	**	**	**
Landownership	**	**	**	**
Land market analysis	**	**	**	**
Population statistics	**	**	**	**
Street maintenance		**	**	
Taxation	**	**		
Utility management		**	**	

Many land data elements are parcel specific and can in theory be collected and collated by a single agency such as a Cadastral Authority. In many countries it is not practical for all the gathering of data to be so centralized since the records have traditionally been compiled separately by different agencies. With modern computer networking it is now possible for these data to be linked into an apparently seamless whole. So-called 'hub systems' contain separate files for groups of topics that can be linked together by using a unique parcel reference number (UPRN) to identify each land or property unit (see Figure 8.4). Examples of a UPRN include a coded version of the parcel's post office address or its cadastral reference number. The total information for one parcel can then be compiled by a series of inquiries to the different organizations gathering the data rather than by retaining all the information on one record.

The advantages of operating in such a way are partly technical inasmuch as a system that is too big and complex tends to generate its own special data management problems. They are also partly administrative in that specialists in one area, such as valuation or public housing administration, can be responsible for the collection, collation, and updating of data within their own field of expertise without interfering with other parts of the system. This allows the land information manager to act as overall coordinator without having responsibility for each individual data item in the system.

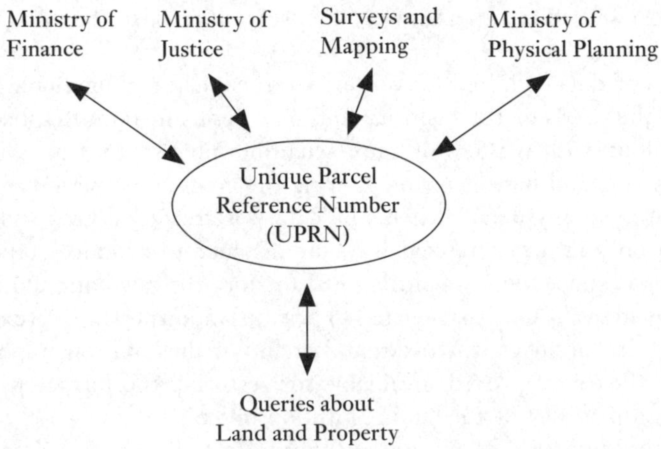

FIG. 8.4 Unique parcel reference numbers

Given the appropriate institutional arrangements, a single authority can take on the responsibility for collecting and coordinating all parcel based information, as has been demonstrated in the case of Service New Brunswick in Canada (Arseneau et al. 1997). In several countries that already have a cadastre based either on fiscal or land use data, the cadastral authority is expanding its role to incorporate additional data and hence to build a more multi-purpose cadastre. This approach allows for a more consistent standard in the quality of the data and can in theory reduce duplication between different authorities. Under a single agency, quality control becomes easier but unless that agency is sufficiently resourced, duplication will arise when other users become dissatisfied with the service that is provided. When data are not in the form required by users or are not easy to access then the users will create their own databases, tailored to their own specific needs. Furthermore, if these users have traditionally compiled their own data sets then they will be reluctant to change their procedures and surrender some of their responsibilities to the centralized agency.

Any broadly based multi-purpose cadastral system must produce information that satisfies its users. This entails discriminating between what it is necessary to know for the immediate purposes of planning and land administration, and what might at some time in the future prove to be useful to know but which at present is not essential. The list of possible features and attributes that may be recorded within any land information system is effectively limitless. If the location of any item is significant, then, technically, it can be included within the system; the key issues then become those of cost and efficiency and what the users rather than the data producers require.

Archives of data collected for posterity rather than for functional purposes may serve the needs of the historian and may be justified on that basis alone; however, along with matters of pure scientific interest, they are difficult to evaluate in cost and benefit terms. During the communist period in east and central Europe many detailed items of data were recorded even though they were used only in a generalized way, for instance to produce statistical information. As such, there was little need to store the raw data. On the other hand, many items of data that related to private landownership were deposited in the historical archives when state ownership of the land was imposed; fifty years later they have proved invaluable for restoring the former patterns of landownership as part of the land restitution process.

One critical test for determining whether to include a specific data category is whether users are prepared to pay the price for its gathering, storage, and updating. The market price for data is likely not to be easy to determine, especially as far as government-held information is concerned. The distinction between what is needed now and what might at some time in the future prove useful is not always clear.

Often the prime beneficiaries of a land information system are the State and parastatal bodies. These organizations may be both producers and users of the system and may not be exposed to full market forces. Undoubtedly the greater the use to which the information is put, the greater will be the benefits and the greater the cost justification. On the other hand, some information may be rarely used but nonetheless worth holding within the system so that it is available at short notice when required—for example in case of an environmental disaster such as extreme flooding or where there is a local emergency, for instance after a railway or an aircraft accident.

In many countries, investment in information technology has been driven by each organization's internal need to improve the management of its own resources. In reality, computerization does not necessarily improve efficiency—there is no benefit in computerizing the mistakes and errors of the past. It can however act as a catalyst to improve existing procedures and has often led to external benefits through the provision of better services to the public.

In many cases, progress has been constrained more by institutional issues than by those of a technical nature. In some countries, especially among the poorer nations, the telecommunications infrastructure is unable to support networking between different sites thus preventing the sharing of data and their treatment as a corporate resource. Such countries tend to have a lack of capital and an excess of labour. Where foreign exchange is in short supply and labour is plentiful and cheap, the social and economic priority may be to maximize local employment and to reject moves towards computerization.

8.3 Internet Applications

Since the 1980s, high-performance workstations connected through local area networks and 'client-server' computing architectures have provided much of the technical infrastructure for land administrators to begin building parcel based land information networks. Recent examples of jurisdiction-wide networking efforts can be found in Australia (Greenwell 1992) and Canada (Arseneau et al. 1997). Most of the early networking projects focused almost exclusively on providing access to secure databases across local or wide area networks using either direct local area networks with interconnected services or high-speed modems (Coleman 1994). Especially since late 1993, however, an alternative model has emerged to address the requirements of access, viewing, and distribution of land data. This approach employs the Internet in combination with standardized graphical user interface software designed to access World Wide Web sites.

Accelerating web popularity has given rise to much of the current interest in the Internet as a means of distributing land data. Hundreds of thousands of websites have appeared since its introduction in 1993, and this growth is continuing to rise exponentially. When used in combination with modern 'browsing' software possessing advanced graphical user interfaces, the Web has emerged as a core communications medium.

The Canadian province of New Brunswick provides an early example of a land administration application of the Internet. In August 1996 Service New Brunswick (SNB) implemented one of the first commercially available on-line land registry systems in the world, a service that provides province-wide access to a series of integrated land data sets. This service, known as the Real Property Information Internet Service (RPIIS), allows clients to access non-confidential, parcel based information residing at a password protected SNB Internet site. The service supports browsing and viewing of maps and map related information over the Internet. Users of this service may search for a property by specifying either a textual or graphical attribute such as a place name or coordinate reference. The software allows users to view and query maps and attributes, select display layers, perform 'point in polygon' analysis, and undertake many more core GIS-type operations. Additional textual and multi-media information can be associated with features on the map.

While the initial web based applications have focused primarily on querying and browsing, the real potential from a land administration perspective is in the fields of electronic commerce (with on-line transactions and updating of both textual and graphical records) and in the promotion of civic participation in the public decision-making processes (MacLauchlan and McLaughlin 1998).

8.4 Costs and Benefits of Computerizing Land Information Systems

Although the use of computers is implicit in all the discussion above, many improvements to existing land administration systems depend more on good organization and management rather than computerization per se. The main advantages of computerization include the:

- physical compacting of data, so that less storage space is required and the data can be more quickly transmitted and hence more rapidly accessed;
- greater facility with which data can be handled and manipulated, allowing much more efficient and effective analysis;
- merging of graphic and attribute data in one set of operations, providing easier understanding for the users; and
- integration of databases so that different data sets can be merged and processed together, creating new information.

As information technology becomes more sophisticated, further opportunities for exploiting data arise, often adding value to the original data sets. This has stimulated improvements in the management of land information and has encouraged the provision of data over wider areas, with more detailed content and greater reliability. Inevitably, as the performance requirements rise, so does the cost of providing the service. Nevertheless the benefits can significantly outweigh the costs (Table 8.2).

TABLE 8.2 Benefits and costs of computerization

Costs	Benefits
Investment in hardware and software	More efficient data manipulation
Maintenance costs	Less duplication
Replacement costs	New forms of data analysis
Data conversion and new data acquisition	Data compacting/lower storage costs
Labour costs	Faster data processing
Overheads	New markets for data

It is difficult to quantify these costs and benefits since many prices are time dependent. At any moment there may be two or more new generations of technology in the laboratories awaiting release as and when the market has grown beyond the present products. When this happens, the price of the old computers and software fall dramatically and the vendors will no longer provide product support. Over recent years, hardware and software production costs have become significantly cheaper and more efficient while the technology has become much more reliable than a decade ago. The benefits of this are however cancelled out by the increased user demands for greater speed, more complex data processing, and the ability to handle greater volumes of data.

Although some estimates of cost are possible, for instance from the overall costs of running a government department as reflected in its budget allocation, quantifying of benefits can prove more difficult. Benefits come from identifying, recording, analysing, and using land related data, examples of which are shown in Table 8.3.

Several attempts have been made to evaluate costs and benefits in hard economic terms but they have had limited success and there have been very few post-implementation studies to confirm whether the predictions became realities. Hence some decision-makers now prefer to approach the investment in information technology on the basis of 'value for money'.

This is a process that is more intuitive than scientific but it can help those who must authorize or underwrite the costs to reach a decision with which they feel comfortable. Each cost and benefit is listed, regardless of whether it is tangible or intangible. Where tangible data are available, the figures are entered on the balance sheet. The approximate cost of an employee per hour (including overheads) is typically around 1/500th part of that person's annual salary. Thus the work of a person on an annual salary of $20,000 per year should be valued at $40 per hour to cover not only the salary but also heating, lighting, office rents, etc. Benefits can then be calculated from estimates of the time saved multiplied by the charge-out rate of those involved. Thus saving five hours' work for a person on $20,000 produces a benefit of $200. Where it is only possible to guess the value in monetary terms (for instance the benefit in money terms of protecting a site of special scientific interest) then the amount entered should be the value with which people feel comfortable (for example US$ 5,000 per year).

In general the costs should be exaggerated and the benefits underestimated. All the figures are then added up and if the benefits still exceed the costs then the decision to invest can be justified with at least a modicum of confidence.

8.5 Paying for Data

Once the decision to invest in an improved land information management system has been made, money for the investment must be found. This may come from grants or loans or from internal sources. In the case of government agencies, taxpayers' money may be provided though, increasingly, governments expect return on their investment through fees charged. The pricing of information goods and services is particularly difficult for governments with little market experience. Some of the factors that influence any charging policy include:

TABLE 8.3 Sources of savings through using GIS/LIS

Archaeological sites—recording and monitoring their locations
Billing customers—ensuring no address is missed
Building-permit allocation, control, and monitoring
Central government reporting—supplying data
Communication—with decision-makers and with the public
Competitive tenders—for instance for grounds maintenance contracts
Compulsory purchase orders—where land is required for municipal purposes
Demographic analysis—including the analysis of socio-economic data
Development control—planning and enforcement
Emergency planning—in case of a major disaster
Emergency services management
Environmental control and enforcement
Environmental impact assessment
Financial control
Finding sites for new developments
Fire services management and checks on buildings
Health and safety analysis and control
Highway maintenance
Land and property taxes
Landownership records
Land searches for local land charges
Land use management
Linkages to other data sets
Managing building assets
Management of mineral resources
Monitoring energy use
Monitoring the land market
Physical planning and resource management
Police management and crime analysis
Pollution monitoring and control
Population forecasting
Pupil data for school resource planning
Resource optimization in general
Road traffic orders and traffic accident analysis
Route planning
Social services management
Street light maintenance
Targeting mail shots
Transport planning in general
Tree preservation orders
Utility management (gas, water, electricity, telecommunications)
Valuation and property assessment

- existing statutory requirements for the charging of fees;
- existing policies and practices for the recovery of costs;
- applications of the concept of a social good and public benefit;
- commercial objectives of individual participating agencies;
- government policies on cost recovery and user-pay principles;
- freedom of information policies; and
- uses to which the data will be put (ALIC 1990).

Pricing may be based on production costs, on market tolerance, or on savings to users. In theory it should be possible to know the unit cost of producing any piece of data though the ability of some agencies to discriminate between different data types and activities may make this very difficult. A further problem is that, given the cost of production for instance of a cadastral plan, there is still a need to know the number of users of that product before a fair estimate of its sale price can be made. The cost of printing additional copies is easy to estimate but the cost of the data contained in the map needs to be divided by the number of maps to be produced.

Many products are priced at the level that the market will bear. In the case of land and property information, such an approach may discriminate in favour of the rich, who are willing to pay more for access to the data than other sectors of the community. A similar difficulty arises when setting the price at the level at which it becomes cheaper to buy the product than for potential users to gather the data themselves or risk having to take decisions without the necessary information.

Charging for data differs from charging for physical commodities since information:

- cannot be consumed—however many times the name 'Washington, DC' is used it will not wear out whereas the more that one uses any physical object, the more wear and tear it will suffer and it will eventually cease to be functional.
- is not transferable, since any information that person A gives to person B can be sold to person C and still retained by person A.
- is indivisible since division changes its meaning—in the case of 'Washington, DC' there is more than one Washington while the letters DC have many possible meanings depending on the context. Thus splitting the name changes its meaning. Conversely, by combining data sets new ideas and meaning are created—the word 'white' has many applications while the word 'house' describes many types of construction but their combination as the 'White House' has a very specific meaning for many people.
- increases in value with use—the more that information is used, the more people will benefit.

- is easily copied—making a copy of a motor car is far more complex than copying a piece of text.

The increasing use of Internet-based data distribution models will probably see many existing users (and even larger groups of new customers) moving from bulk-ordering practices to a more transaction-oriented approach to the delivery of land data. People will access the data as and when they want to use it. Early experiences by the US Fish and Wildlife Service in the United States indicated that providing Internet access to digital datasets can dramatically increase the number of requests for such products by both regular customers and impulse-driven 'data grazers' (Coleman and McLaughlin 1997). Similar experiences by Service New Brunswick with their web based service point to the same increase in customer interest (Arseneau et al. 1997).

Such experiences show that:

> better access can lead to wider usage of the datasets in question. However, existing data pricing philosophies and fee structures—often developed on per-file, volume-discount bases—may be ill-equipped to deal with an environment in which users want to browse through information holdings before selecting an area of interest which may represent only a small portion of an existing file or a clipped combination of several files. (Coleman and McLaughlin 1997)

The growth in interest in Internet-based activities heightens the concerns about the prospect of information 'haves' and 'have-nots'. Current statistics suggest that the on-line user community is largely made up of younger professionals and technical specialists. While the situation is slowly changing, many current users of traditional hardcopy records and maps may be either unaware or intimidated by the growing availability and use of on-line data products.

This situation is even more acute in less developed jurisdictions, where land administration records and property maps have been traditionally in either short supply or unavailable to the general public. Nevertheless, there is growing availability of spatial data in these countries and many international aid organizations and private companies are contributing to this supply. However, this fact is of little use in countries where literacy levels are low, few computers are available, or the communications and power supply infrastructure is unreliable. There are no easy answers or quick solutions to this problem, but 'Unless public and private data suppliers continue to provide data in a variety of electronic and hardcopy forms for the foreseeable future, the technology and "technology-application" gaps between developed countries and emerging nations will increase rather than diminish' (Coleman and McLaughlin 1997).

8.6 Intellectual Property Rights in Land and Property Information

Copying is facilitated by computerization since the creation of multiple copies of any text or graphic in digital form is now so easy and so cheap. When printing presses were first invented, copyright laws were developed to protect the printers rather than the authors. Only in recent times has there been concern about the intellectual property rights, invasion of privacy, and product liability arising from the wide availability of information.

Copyright is a property right that exists in creative works such as literature, drama, music, painting, and sculpture. It is designed to protect the interests of the data producers. Copyright law can be applied to many types of land and property information including maps and tables. Within the European Union, new legislation has been introduced to protect data in databases since under earlier law protection only extended to the way in which the data were compiled and not to the data themselves.

Copyright law allows data producers to protect their investment. Under such legislation, copying means reproducing works in any material form, including the storing of work in any medium by electronic means. It covers data and compilations of other people's work as well as computer programmes, making copies in three dimensions, and any adaptation of original material, including copies that are transient to some other use.

A high level of investment has taken place in the creation of topographic and cadastral maps and in the gathering and storage of land related data. Given the pressures on many government agencies to recover their costs, it is inevitable that government agencies seek to protect their investment through copyright laws. There is an argument that since most of the gathering of the data was paid for through taxpayers' money, the data should be made available to the public for free or, at maximum, at the cost of handling the data to make them accessible. There is a contrary argument that the costs of maintaining the system should be paid for by present-day users. The economic pressures are not directed at recouping past investment, but rather at maintaining or enhancing the present volume and quality of data. Hence the investment must be protected and costs recovered through copyright fees.

In practice, copyright laws are difficult to enforce. Technically, every time a user studies a printed paper version of a map, no copy is made; conversely, to examine a digital map necessitates copying the data from a computer storage disc and transforming those data into an image for display on a screen. Thus every time a digital map is displayed, a copy is made but every time a paper map is taken out of a drawer or off a shelf, there is no copying. As a consequence, the dynamics of copyright enforcement are resulting in digital data providers leasing their data rather than selling them. By this means, there is a right to use the data, the copyright fees being based on estimates of the extent of use. Copyright in effect becomes a use right rather than an ownership right.

Other intellectual property laws that may affect land related information include data protection laws that are designed to protect the privacy of individuals. Under such laws, citizens have the right to know what data are held about them. The definition of personal data is broad and includes data of any type that may be related to an individual. Thus the fact that a house belongs to an individual is a piece of personal information, the use of which must be declared at the time when the data are gathered. The use of the data must be registered and the individual must be informed about how the information will be used. As with copyright, enforcing the law is often difficult but data providers run the risk of prosecution if they fail to abide by such regulations.

Not all countries have such legislation although, for example, those former communist countries currently applying to join the European Union (EU) are under obligation to bring their legislation into line with EU Directives.

References

Arseneau, B., Kearney, A., Coleman, D. J., and Quek, S. H. (1997), 'Internet Access to Real Property Information', *Proceedings of the GIS'97 Conference*, Vancouver, British Columbia.

Australian Land Information Council (ALIC) (1990), 'Charging for Information', Issues Paper No. 3. Belconnen: Paper prepared for the Australian Land Information Council.

Coleman, D. J. (1994), 'Geographic Information Systems Performance in a Broadband Communications Environment', Hobart: unpublished Ph.D. Dissertation, Department of Surveying and Spatial Information Science, University of Tasmania.

——and McLaughlin, J. (1997), 'Information Access and Network Usage in the Emerging Spatial Information Marketplace', *Journal of Urban and Regional Information Systems Association*, 9/1.

Dale, P. F., and McLaughlin, J. D. (1988), *Land Information Management* (Oxford: Clarendon Press).

Greenwell, J. (1992), 'Communications Issues for Land Information Systems', Seminar Notes and Materials, 1992 Workshop on Networking Spatial Information Systems. Hobart: Centre for Spatial Information Studies, University of Tasmania.

MacLauchlan, W., and McLaughlin, J. D. (1998), 'Environmental Democracy and Stewardship through Land Resource and Information: The Case of the Land

Gazette', in M. Barry (ed.), *Proceedings of the International Conference on Land Tenure in the Developing World*, Cape Town.

Wyatt, P. (1998), 'Using a Geospatial Property Information System to Analyse the Effect of Accessibility on Retail Property Value', University of the West of England, Bristol, Faculty of the Built Environment Working Paper 42.

9

Managing the Land Administration Process

9.1 Management and Management Information

Technology rarely poses the major concerns in any effort to build and sustain an effective land administration infrastructure. Rather, the core challenges tend to be related to management issues, as will be discussed in this chapter.

The Management Process

Management was described in our earlier book as the art and science of making decisions in support of certain perceived objectives. Like politics, it is the art of achieving the possible, the ability to get things done. Management skills are needed in order to implement policy decisions and to meet the objectives set for any organization. Good management seeks to do this in an optimum fashion—perfect solutions never arise. Management entails extrapolating trends from a limited range of facts—sufficient information is never available for decisions to be made with certainty as to their outcome.

Better information will however bring about a better understanding of any system and hence create the possibility for its better operation. Information is needed to:

1. monitor what is going on so that areas where decisions need to be made can be identified;
2. help evaluate alternative strategies for dealing with the problems or opportunities that have been identified;
3. assist in selecting the right course of action; and
4. facilitate the implementation of whatever has been decided.

Management operates at all levels from the personal to the institutional; for instance all individuals need to practise self-management in order to achieve their optimum personal goals. More particularly, management within an organization operates at three key levels—strategic planning, management control, and operational control. Strategic planning is the process whereby decisions are made on an organization's objectives, including where the organization should position itself to cope with future trends and markets. Operational control involves the day-to-day processes of ensuring that routine tasks are carried out efficiently and effectively. Management control is the interface between these two, ensuring that adequate resources are secured, either in terms of people, money, or equipment, to achieve the organization's mission and objectives.

Management Information

The information required by an operations manager differs from that required by someone concerned with strategic planning. Operations managers must meet the immediate needs of their clients and employees and will therefore need very focused and accurate data obtained largely from within their organization's own records system. Strategic planners on the other hand will be looking at general trends and anticipating changes in the market; they will be concerned with the past, the present, and the future and will be more concerned with what is taking place in the outside world than what is currently happening to the organization.

TABLE 9.1 Information needs of managers

Data quality characteristic	Needed by	
	Operational managers	Strategic planners
Accuracy	high	low
Amount of detail	specific	aggregated
Data source	mainly internal	external
Scope	narrow/focused	very broad
Frequency of use	often (e.g. daily)	infrequent
Time scale	historical (archives)	future
Currency	up to date	historical

The provision of appropriate information for such managers must also take into account the fact that data may be interpreted in different ways. Different individuals will draw different conclusions from the same observations depending on their background, understanding, and the context within which they turn data into information. More particularly the fitness for purpose of any data depends on:

1. the context in which the data are used—the same facts can be interpreted in different ways depending on the circumstances;
2. the data source, that is from where the data have come—since this will determine the reasons why the data were originally collected;

3. the content and nature of the data—graphical images for example have a different impact from those verbally described;
4. the personal characteristics of those who are making decisions—individuals tend to interpret data in the framework of their own experience;
5. the learning capacity of the user—especially where words and images are used in accordance with a defined classification system;
6. the rules and regulations affecting access to the data—some data may not be able to be released for security or confidentiality reasons; and
7. the time at which the data are accessed, as this may affect their context and current accuracy.

Because many organizations responsible for land administration are in the government sector and have traditionally been less exposed to market forces than private sector organizations, the availability of management statistics on productivity and data on the true cost of the service are often severely limited. In many countries, productivity has been low and staff have not been strongly motivated. On occasion, the real cost of producing a paper description of a land parcel has, for example, been higher than the market price of the land itself. This may be because inappropriate standards of measurement have been sought or as a result of poor human and technical resource management and control.

Significant cost savings may be possible through the use of improved management techniques. A first step on the road to improvement is to acquire better management information and to monitor what has been going on within an organization. All the existing procedures associated with the day-to-day handling of land information need to be carefully documented through the use of quality assurance techniques.

Quality Assurance

Quality is a measure of how well a product or service meets the needs of those who use it. Quality Assurance, commonly known as QA, is a regulatory mechanism whereby an organization follows certain procedures that should, in theory at least, reassure a client that what is being provided meets his or her needs. QA is concerned with all those features and characteristics affecting an organization's ability to meet its objectives. In all processes of production, there should be quality controls to check that standards are being maintained. Some checks are built automatically into the procedures that must be followed; for instance, checking the closure of a surveyor's traverse or that the three measured angles of a triangle add up to 180 degrees. These measures provide automatic quality control. Many quality controls are however based on a strategy of sampling, testing only part of the production process—an example of risk management.

Quality assurance ensures that adequate quality controls are carried out. It is a process in which planned and systematic actions are conducted in a way that provides confidence that a product or service will satisfy given user requirements. Quality does not necessarily mean excellence; rather it means that which is necessary and sufficient to meets the needs of the client. Thus if residents on a housing estate wish to know how far it is to the nearest grocery shop, it is not necessary to give them the answer to the nearest metre let alone centimetre or millimetre. Quality must be seen in the context of every user's needs, ability to pay, and understanding of what the product can and cannot do.

Quality assurance may be achieved by monitoring all procedures involved in the manufacture of a product or the provision of a service. It may be carried out internally or by a third party, that is an external body that scrutinizes the documentation of all the relevant procedures. Some companies believe that their name and reputation is of such standing that no external scrutiny is needed—they trade on their reputation and if that is tarnished by bad work, then market forces will drive them out of business. Such firms argue that it is in their own interest to ensure quality and that formal quality assurance is unnecessary. Others look for guidance from an outside body that will give them accreditation so that all clients can be reassured that quality procedures have been scrutinized and are being followed in accordance, for example, with the international standard series ISO 9000.

Quality assurance is a management control system in which all activities are written down and clear lines of responsibility for their execution are determined. It is concerned with accountability such that at every stage of an operation, all individuals know what their functions are and some individual has direct responsibility. In simple terms, QA means that if anything goes wrong, it is clear who is to blame, that each person has been told his or her responsibilities in writing, and that all tasks have been fully documented.

Current management fashion recommends that each organization should have a mission statement defining what it is in business to achieve. From this its aims and objectives can be derived, an aim being something intangible like excellence while an objective is something that is measurable. Objectives can take the form of specifying a particular level of turnover that can be shown on a balance sheet, or requiring 90 per cent of all letters received by an organization to be acknowledged within three days. The tasks that flow from these aims and objectives must then be identified and rules and procedures set in place to ensure that they are achieved. All this should be documented and every individual should then be given a clear and concise job specification.

The process of documenting an organization's procedures is time consuming and, depending on the size of the organization, can take up to or more than one year. Although it is an expensive process, many companies that have achieved accreditation agree that it improves the efficiency of their organization, eliminates duplication of activities and effort, and ensures that staff are more conscious of their responsibilities.

The key element to QA is auditing. It should at all times be possible for an outside agency to investigate what is being done, why it is being done, and by whom it is being done. Each person responsible for doing a task must know and understand what is required and be capable of delivering what has been specified.

Total Quality Management

Total Quality Management (TQM) is a way of improving the quality of what is delivered by encouraging the total commitment and participation of a workforce. It requires rigorous attention to all processes within an organization from input through processing to the final delivery of products and services so that the needs of each and every customer are satisfied. QA is concerned largely with internal procedures as to how this may be achieved—indeed it has been criticized as being over-dependent on documentation to the detriment of individual responsibility and initiative, and over-concerned with short-term needs without considering longer-term strategies and responses to changing market requirements. TQM is a management technique that transcends mere QA.

According to Stirling (1993) and others, the philosophy behind TQM is that:

1. rather than concentrate on improving the quality of the product, one should concentrate on improving the processes that create it, and build quality in at the design stage;
2. prevention is essential since quality cannot be inspected as it is dependent on the satisfaction of the end user, which can only be established after the product or service has been created;
3. all action must be customer driven;
4. quality improvement requires the commitment of top management—it has been estimated that management, not the workforce, is responsible for over 80 per cent of quality problems;
5. performance quality measures must be established so that all individuals know what is expected of them;
6. quality improvement requires involvement and commitment of everyone in the organization—quality is an organizational culture that has to become a way of life;
7. both initial and ongoing training are essential—organizations must invest in their workforce and ensure that they undertake continuing professional development;

8. there are no short cuts to quality since quality is a continually moving target.

The key issue is the commitment of the top managers to total quality management. These managers must create policies on quality and allocate sufficient and appropriate resources to achieve the necessary organizational objectives. Management is responsible for motivating personnel, for providing them with all necessary training and for creating opportunities for relevant self-development. It also has prime responsibility for communication within the organization so that all staff are informed of what is expected of them and the reasons why quality is so important. Conversely, managers must listen to their workforce so that they know and understand the needs of those on the shop floor and in middle management posts. They must also listen to their clients and the market place.

According to the UK Department of Trade and Industry (DTI 1991), TQM helps companies to focus clearly on the needs of their markets and to achieve top quality performance in all areas, not just in products or services. It encourages organizations to operate simple procedures that are both necessary and sufficient for the achievement of a quality performance and to eliminate non-productive activities and waste. Embarking on QA without going the whole way to Total Quality Management is in general regarded as a mistake. Every client needs and expects quality and the only effective way to ensure this is through TQM. QA is a step along the path but it is not a goal other than in the short term.

Risk Management

One component of TQM is risk management, which may be defined as the process used to manage exposure to risk. There is a spectrum of techniques for minimizing risk (Green et al. 1992; Vaughan 1992), including:

1. Risk avoidance. By consciously not engaging in the action that gives rise to the risk, the chance of loss is reduced to virtually zero.
2. Reduction of the frequency of loss. The installation of physical security measures in a registry office, for example, can reduce the chance of a loss occurring.
3. Reduction of the severity of loss. While reducing the frequency of loss may have significant results, some losses will inevitably still occur. Hence other measures should be designed to reduce the magnitude of these after they have occurred—for instance by installing a sprinkler system in a land registry to allow a fire to be extinguished immediately.
4. Sharing the risk. Some risks may be shared by transferring them to

insurance companies; by pooling large numbers of similar exposures to risk, losses sustained by a few can be shared by the entire pool.

5. Transferring the risk. Risk may be shifted to others through a variety of means, for instance by incorporation into a business with limited liability. Other examples include guarantees or contracting out the liability to others such as professional surveyors and lawyers. New Brunswick, for example, has recently developed an innovative approach to sharing the responsibility for managing risks in its registry system with the legal and surveying professions.
6. Retaining the risk. Strategies may be developed for managing those risks that have not been avoided, reduced, or transferred. Active or voluntary retention means that the risk holder is consciously aware of the risk and deliberately plans to retain it. Passive retention results when certain risks are unknowingly retained because of ignorance, etc. Risk retention is often appropriate for high-frequency, low-severity risks where potential losses are relatively small.

A recent and intensive examination of risk management as it applies to the land registration function can be found in Palmer (1996). Consideration of risk management is crucial if the costs of cadastral surveying are to be reduced. There need to be appropriate standards in terms of levels of security, information content, and boundary precision that reflect more effectively the economic and social requirements of the community. High-cost, high-precision surveys are not an option in most countries, especially those that are less developed or are in economic transition. The priority must be to meet the needs of the time, leaving as an option the upgrading of the system at a later date when market conditions dictate (McLaughlin and Palmer 1996).

9.2 Re-engineering Land Administration Systems

One of the significant trends in land administration has been the move towards a more commercial and market driven approach to the delivery of government services, with some systems becoming fully self-financing. This is affecting the operations of land registries, cadastral agencies, and national mapping organizations. There is a general acceptance that the formalization of land titles requires government subsidy but that the ongoing processes of maintenance and update of the land records should be subject to full cost recovery. There has been good progress towards this in the operation of land registries where text data are handled and the user community can be charged for the delivery of the service. In the business of surveying and mapping, full cost recovery has been much more difficult to achieve; subdivision work can be carried out at cost by either the public or the private sector but almost all national mapping agencies receive government subsidies. Rhind (1997) has reviewed progress towards the commercialization of national mapping.

The need for greater cost recovery has also led to the restructuring of many government and quasi-government organizations and a redefinition of their roles, especially in relation to the private sector. The process is often referred to as re-engineering, a term used in the management consultancy industry to define the 'radical redesign of a company's processes, organization and culture to achieve a quantum leap in performance'. The term received widespread recognition with the publication of *Re-engineering the Corporation* by Hammer and Champy (1993).

Re-engineering is one of many so-called management 'technologies' that have emerged over the years to assist companies in meeting more effectively their corporate objectives such as increased revenue, greater market share and/or profit, lower costs, and improved services.

Re-engineering is distinguished from other management techniques in terms of its radical nature. It is neither restructuring, nor total quality management, nor performance improvement, but rather a sophisticated procedure producing top-to-bottom organizational transformation. An extensive literature has evolved assessing the value of re-engineering (see for example Hall and Rosenthal 1993). The process has many critics who argue that it is just another management fad without much substance, or that it is just a euphemism for corporate downsizing and the laying off of employees. In the land administration arena, there has been some downsizing as a result of computerization but the processes of data conversion and quality improvement have absorbed some of the spare capacity that often existed. Furthermore, reductions in government staff have often been matched by a growth in the private sector, who have taken on new work.

In practice, much of the drive for re-engineering is coming from computerization and as a response to the information technology industry. In many mapping agencies, for example, investment in high technology has resulted in downsizing and internal reorganization. There has however been a growth in the private sector since, as a critical mass of data become available in digital form, new industries have begun to spring up exploiting those data. The people involved in these new industries are not necessarily the same as those made redundant by internal reforms within the public sector but the net effect is positive.

The processes of re-engineering, total quality management, and other approaches (such as 'management by objectives' and 'process innovation') emphasize different methodologies and priorities. Nevertheless, they all advocate activities such as planning, analysis and monitoring, informed decision-making, and the identification of an organization's missions and objectives as part of the process of improving products and services. Underlying all of these approaches is the assumption that the information infrastructure necessary to perform key activities is present, that is there are proper management information systems in place.

9.3 Technical Reforms

Part of the drive towards re-engineering has come from the opportunities created by new technologies. Technical reforms have concentrated on advances in:

- information technologies for processing and managing property information;
- communication technologies for providing effective access to information and services; and
- data gathering technologies, especially in surveying and mapping.

The most significant change in land administration over the last decade has been the extent of computerization of the land registries. The objective of computerization has been primarily to meet internal requirements for more efficient data storage, more rapid information retrieval, and greater ease in updating the records; only in a second phase have the benefits to the public begun to appear. Separate initiatives have taken place in the agencies responsible for cadastral and topographic mapping and these have often been driven more by the technology than by the need to provide a better service.

Computerization of land registries is not new. Computers were introduced into the land administration field in the early 1960s, first in support of the property valuation and surveying and mapping functions. Automation of land registry systems began around 1970, the most often referenced pioneers of the period being Sweden, New South Wales in Australia, and New Brunswick in Canada. Computers were introduced to support the accounting function, to assist in the examination process and especially in reviewing survey plans, to prepare automated indexes, and finally to develop automated databases.

Computerization also helps to provide back-up facilities in case of disaster. Microfilm technology has in the past been used to provide a secure back-up record of documents in the registry; this function is now often managed using optical scanning technology. Some countries have failed to maintain disaster copies of their records because of the high cost; they may have assessed the risk of fire or natural disaster and concluded it worth taking, but more commonly and especially in less developed countries they simply have had no funds available. When eventually they computerize their records, as many are already beginning to do, the cost of retaining back-up copies of the registers should be minimal. In Sweden where the registry records are all fully computerized, there will in future only be electronic forms of back-up.

The most technically advanced registries now provide on-line computer access, increasingly via the Internet. They provide direct access to property records including information about the legal owners, their addresses, current legal interests and charges against property, legal descriptions, and references to off-line digital storage for more detailed documentary material. Complex security arrangements have had to be put in place to prevent hackers breaking into the systems and altering the records.

The most recent information technology developments in the land administration field have included (i) the construction of modular land records management software systems (for example, the NovaLis system from Canada); (ii) the development of Internet browser tools for both textual and spatial querying of property information (several jurisdictions, such as New Brunswick, now provide complete on-line access via the Internet to the property records); and (iii) the use of simple field-operated devices for gathering information for valuation and digital mapping.

There have also been significant advances in understanding how to employ these advanced technologies effectively. As Burns et al. noted: 'Technology has a vital role to play in land titling but it has to be looked at within the overall objective of establishing a land administration system. Decisions on technology made in land titling can have a major impact on the successful integration of the records into the land administration system.' Where it was shown that new technology could overcome a production bottleneck, and that the new technology was sustainable, then it was carefully integrated into the agency. Associated with this introduction was a carefully planned programme of in-country and overseas training. However, of equal significance to the overall success of the projects has been the review of existing manual procedures such as the simplification of a dealings form, or the streamlining of an administrative procedure (Burns et al. 1996).

The main technologies that have affected cadastral surveying have been the global position systems (GPS), digital mapping, and geographic information systems (GIS). Each of these technologies is beginning to influence the way that data are gathered and the types of data that are being collected. Another technology, high-resolution remote sensing, has so far had only marginal impact on the cadastre; however this is likely to change in the near future.

GPS is already used in cadastral surveying, mainly for the establishment of control frameworks. The use of GPS, together with digital cadastral mapping, is in line with general developments in surveying and mapping. The use of GIS in the cadastre is however only just beginning since there has been insufficient data in digital form to justify their implementation. Restrictions on general access to cadastral databases further discourage their use. These restrictions, while allowing specific information about individual land parcels to be retrieved, prevent a general sweep of the whole of the land titles database for reasons of confidentiality. This for example prevents detailed comparisons being made between apparently similar properties that have different market values, an analysis that GIS technology could in theory easily make. GIS has therefore yet to prove itself in the field of land administration.

Barnes and Eckl (1996) reviewed technological innovation in land administration. They suggest that all new technologies should be justifiable in terms of:

- speed (they must significantly outperform current approaches);
- cost (they must significantly reduce current unit survey costs);
- suitability (they must be within the financial reach of local surveyors and within their range of skills to operate);
- appropriate accuracy (they must match real needs);
- simplicity of field operations (the data collection must be simple enough to allow for many different field conditions).

New surveying technologies have reawakened the debate about appropriate standards of precision in measurement and how these standards should evolve over time. They have also moved the emphasis for surveyors from measurement to the management of the technology. Although GPS technology in particular is having an increasingly important impact on the measurement of parcel boundaries and on the construction of property maps, the greatest impact on land administration has come from information-handling technologies rather than from surveying.

9.4 Project Management Methodologies

Based upon extensive experience in countries such as Thailand (Burns et al. 1996; Rattanabirabongse et al. 1998), fresh ideas and project management methodologies are under construction around the world. One example is the Land Information Centre Project Management Methodology (LICPMM) developed by the Land Information Centre, Government of New South Wales in cooperation with a series of Chinese partners (Grant et al. 1996).

LICPMM has been developed and tested in the Zhuhai Municipal Government of Guangdong Province in China. While the methodology has been specifically designed for the land information component of a land administration project, much of it is of general applicability. The methodology has had to contend with the constraints imposed by different cultural backgrounds and communication barriers (not only different languages but also long distances), multiple objectives of the participating agencies, and limited access to key staff.

LICPMM was employed to:

1. define the project management structure, responsibilities of each participating organization, reporting and dispute resolution procedures;
2. break down the tasks into a set of sub-tasks, deliverables, and deadlines, and assign individual team members to specific sub-tasks;
3. develop rigid quality assurance and documentation procedures for each stage and for project review procedures.

A three-tier project management structure is incorporated in the methodology for effective coordination with:

1. A steering committee that oversees all aspects of the project, and is composed of the principals from the key participating organizations. The committee is responsible for executing agreements and ensuring that relevant activities are effectively performed by the appropriate organization.
2. A project management team, consisting of the relevant project managers from each organization, that is responsible for the day-to-day operations of the project and for periodic reporting to the steering committee.
3. A group of key technical staff who are nominated by each organization to carry out specialized tasks.

In addition, there are staff responsible for managing the quality assurance programme.

LICPMM incorporates more than 200 sub-tasks, assigns individuals to specific tasks, and incorporates specified deadlines and reporting mechanisms. The Quality Assurance component comprises a number of standards and procedures, and addresses the issues of: delegation of responsibilities; design and document control procedures; configuration management; process control, inspection and testing, and project review procedures; documentation and records standards; dispute resolution procedures.

An even more ambitious methodology has been under development in Peru for more than a decade. Known as PROFORM (McLaughlin and de Soto 1994), it attempts to integrate all aspects of a titling and registration project (from the political level, through community participation, to field titling campaigns, adjudication and registration processes, institutional strengthening, training, etc.). The involvement of the community in the design and implementation of this project has been of particular significance—the approach has been very much bottom up rather than top down. The new registration system was designed to cater for the needs of poor residents of informal settlements and, increasingly, institutions that grant credit to them. By re-engineering the registry processes, the new system significantly cut the time and costs involved in transforming informal holdings into formal property. Re-engineering also

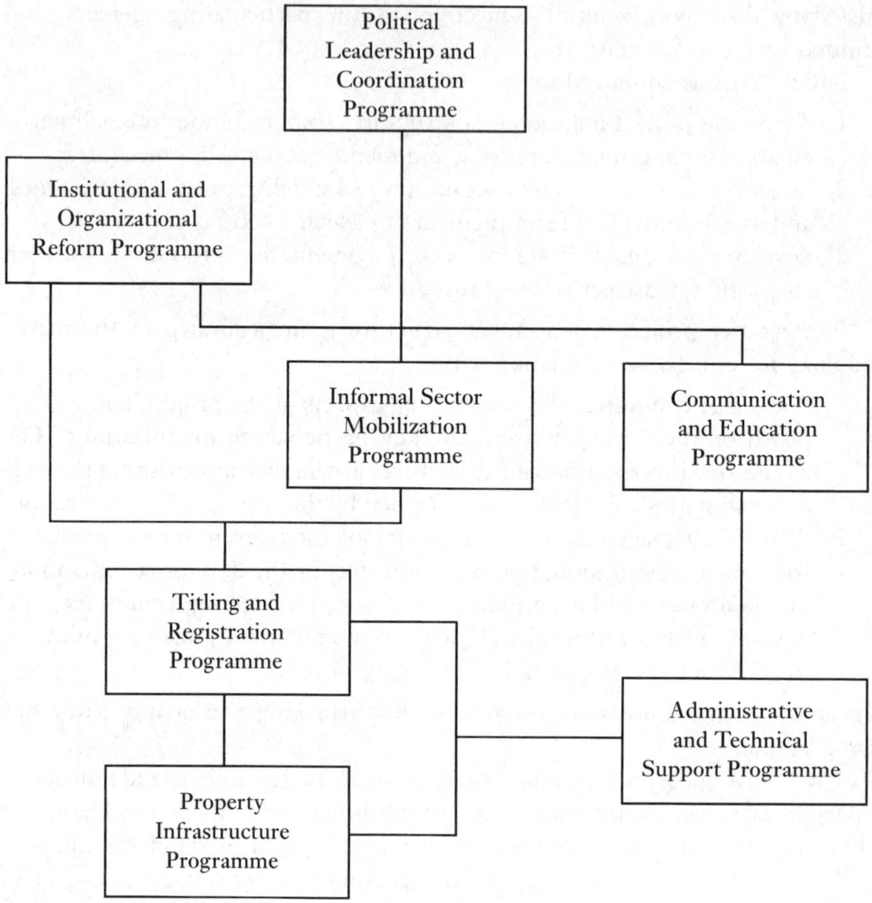

FIG. 9.1 PROFORM methodology

created incentives to keep this newly formal property in the registry system by reducing costs of subsequent transactions such as registering sales and mortgages.

The drive for reform came from the Institute of Liberal Democracy (ILD) which used its knowledge of informal communities to mobilize residents, thus allowing property information to be collected efficiently. Since information was collected openly from all community members at the same time, people had the opportunity to scrutinize claims made by others.

The ILD's legal and surveying verifiers inspected information collected from the informal communities along with existing legal documents obtained from government archives. By reconciling discrepancies and redrafting documents, the ILD helped agencies to produce high-quality titles and plans. Because the issue of quality control was addressed at an early stage, these high-quality titles could be registered in a new registry system, the Registro Predial, without having to follow extensive security precautions.

The reform also reduces costs of registering formal transactions, thus providing an incentive for newly formalized properties to remain formal. It seeks to minimize cumbersome and costly requirements of transfer like certificates from municipalities regarding payment of taxes and from notaries. Short, simple standard transaction forms reduce ambiguity encountered in older multi-page, unstructured documents. The computerized parcel based registry further reduces information asymmetry by enabling ownership information to be determined in a matter of hours as opposed to months in the Registro de la Propiedad Inmueble.

Furthermore, property transactions can be completed in a week in the Registro Predial compared to 6–9 months in the other registry. In addition to simplifying registration procedures and reducing their costs, the reform increases the level of security enjoyed by lenders who provide loans secured by mortgages. A number of commercial, state, and non-governmental organizations have extended credit secured by mortgages registered in the Registro Predial.

The case of Peru emphasizes the importance not only of providing access to registered title but also of doing so efficiently. Credit in developing countries is often given in small amounts (under $1,000) and for a short time (less than a year) since a person's income (a measure of their ability to repay the loan) is as much a factor as any collateral offered. If the registry system is inefficient, the costs of registering a mortgage will be a high percentage of the loan. Furthermore, delays in the registration process may result in the loan being repaid before the mortgage has been registered. As a system grows more efficient, it becomes cost effective to register smaller mortgages, and thereby also increases the incentive that future changes in ownership will be registered.

By providing an effective formalization and registration process, the Peruvian reform has stimulated growth. As described earlier, homes in informal settlements start off as flimsy dwellings made of straw matting; in the past, because credit was scarce, residents waited years to save the cash needed to buy bricks, cement, reinforced steel, and other supplies. One lender, a supplier of building materials, now provides credit secured through registered mortgages to enable newly formalized owners to improve their homes today. Recognizing the potential business arising through widespread registered ownership, that lender is now financing the creation of more formalized property through a cooperative agreement with the Registro Predial.

References

Barnes, G., and Eckl, M. (1996), 'Pioneering a GPS Methodology for Cadastral Surveying: Experiences in Albania and Belize', Orlando, Florida: paper presented to the International Conference on Land Tenure and Administration, November.

Burns, A., Eddington, R., and Nettle, K. (1996), 'Land Titling Experience in Asia', Orlando, Florida: paper presented to the International Conference on Land Tenure and Administration. November.

DTI (1991), *Total Quality Management: A Practical Approach* (London: Department of Trade and Industry).

Grant, D., Zheng, Y., and Li Hua (1996), 'LICPMM: A Blend of Chinese Culture with Western Project Management Philosophy', in P. Zwart (ed.), *Proceedings of AURISA 96* (Hobart: AURISA).

Green, M., Trieschmann, J., and Gustavson, S. (1992), *Risk and Insurance* (Cincinnati: South Western Publishing Co.).

Hall, J., and Rosenthal, J. (1993), 'How to Make Re-engineering Really Work', *Harvard Business Review*, 71/6.

Hammer, M., and Champy, J. (1993), *Re-engineering the Corporation* (New York: Harper Business).

McLaughlin, J., and de Soto, H. (1994), 'Property Formalization: The PROFORM Solution', *Geomatica*, 48/4.

——and Palmer, D. (1996), 'Land Registration and Development', *ITC Journal*.

Palmer, D. (1996), 'Incentive-Based Maintenance of Land Registration Systems', Gainesville: Ph.D. Dissertation, University of Florida.

Rattanabirabongse, V., Eddington, R., Burns, A., and Nettle, K. (1998), 'The Thailand Land Titling Project—Thirteen Years of Experience', *Land Use Policy*, 15/1.

Rhind, D. (ed.) (1997), *Framework for the World* (Cambridge: Geoinformation International).

Stirling R. (1993), 'Quality: A Much Misunderstood Issue', *Surveying World*, 1/2.

Vaughan, E. (1992), 'Fundamentals of Risk and Insurance', New York: John Wiley & Sons.

10

Policy Issues in Land Administration

10.1 Reinventing Government

Land administration strategies and processes need to be structured within a broad policy framework, the shape of which will depend on the jurisdiction concerned. A common thread between systems will be the promotion of economic development, social justice and equity, political stability, and environmentally sustainable development. The processes of re-engineering, total quality management, and other management reforms discussed in Chapter 9 were originally designed for use in the private sector so that organizations could respond better to the demands of the market place. More recently, they have increasingly been adopted by public sector administrators who have been forced to respond to the market oriented approach and hence have been required to upgrade land administration systems.

In the United States the processes of re-engineering have been packaged under such labels as 'entrepreneurial government' and 'reinventing government' and were addressed in the National Performance Review (known as the Gore Commission) which had a mandate to 're invent and to re-invigorate the entire national government'. The ideas were picked up by many other governments—from Australia to the UK (Butler 1994), the Netherlands to New Zealand, and Singapore to Sweden—regardless of party or ideology.

Although reinventing government means different things to different people, it has generally entailed:

1. restructuring the way government services are organized;
2. developing new strategies and processes for managing government services (for instance, simplifying administrative programmes);
3. empowering the recipients of public services.

As with the private sector, a crucial component of reinventing government has been the effective use of information technology (IT). Governments in general have only recently begun to review their national information strategies and to develop new ways in which they deliver services to citizens and businesses. Over the past few years, IT has changed the way that many people live through the creation of new products and services. Examples include the use of credit and debit cards, the ability to withdraw cash from a 'hole in the wall' even in a foreign country, the mobile phone and fax machine, and access to information on the Internet. Information technology now makes it possible for citizens and businesses to deal directly with government agencies if they so wish (UK Government 1996).

The motive behind the increased use of IT in government has in part been to improve the delivery of services; at the same time it has often been (overtly or otherwise) part of a strategy to reduce the number of civil servants and hence the costs of government. For more than a decade most developed countries have been working on strategies to cut back on government in the economy, and these have met with some success (with public expenditure as a share of GDP being stabilized if not reduced). These strategies have to some extent been inspired by neo-conservative political ideologies that came to prominence in the 1980s, but they have also been fuelled by:

1. concerns about the size and scope of the public sector and the way that this hampers private sector performance;
2. a growing appreciation of the social costs of higher taxation;
3. concerns about the effect of regulation and public ownership on innovation; and
4. a reassessment of the nature of and need for natural monopolies.

Over the last decade, there have been initiatives in many countries aimed at reducing government intervention and expenditure on the one hand, while improving efficiency in the use of resources in the remaining public sector programmes. Initial efforts to restrict the growth of government took the form of top-down budgetary constraints and the creation of target levels for manpower and operating costs. These have, however, met with only limited success. More recently, reforms have tried to:

- introduce private management practices and to simulate market processes;
- define more sharply the objectives and outputs of government agencies;
- develop productivity measures and provide managerial incentives for reducing costs and improving performance; and
- develop better accounting and management information systems (for example, the introduction of accrual-based accounts in New Zealand was aimed at gauging how efficiently capital was being employed and to prevent output targets being achieved by running down physical or human capital rather by than improving services).

10.2 Privatization

Arguably the most fundamental reform within the public sector has been privatization. Here the term 'privatization' refers to any shift in activity from the public to the private sector. This might involve merely the introduction of private capital or management expertise into a public sector activity; more typically, however, it involves the transfer of ownership of public enterprises to the private sector.

Privatization policies have been aggressively pursued over the past decade in several jurisdictions around the world. There have been massive privatization programmes in the United Kingdom, Japan, and other major industrial countries, as well as in smaller economies such as New Zealand. They often require a shift in culture and the elimination of deep-seated prejudices held by those who have only known public service and have a deep distrust of the private sector.

As an economic policy, privatization has been promoted on the basis that private ownership and control is more efficient in terms of resource allocation than public ownership. There is a presumption that public and private enterprises have different incentives and hence different efficiency outcomes. Privatization does not necessarily eliminate the monopoly profits, if any, that were enjoyed by public enterprise—for instance the privatization of electricity and telecommunications services has, in a number of countries, created organizations that are profiting from their monopoly positions. In defence of these monopolies, it is sometimes argued that privatization brings under the control of market forces those profits that were previously non-tradeable. Through privatization, an enterprise can gain access to private sector financing and private owners may bring exposure to new markets. It is argued that when privatized, the agency is in general more capable of responding to demand—for instance by shedding surplus staff when the demand for services is low and recruiting additional staff at times of high demand. The government service cannot in general be so responsive to the market.

In the field of land administration, although much work has been 'outsourced' to the private sector, the general policy has been to retain the core agencies and functions within the public sector. At the same time, recognition of the merits of competition has put pressure on such agencies requiring them to be more commercial and to recover most of if not their entire running costs. In a somewhat related area, Rhind (1997) has discussed many of the problems and opportunities that national mapping organizations are facing as governments encourage them to become more self-sufficient. Almost all of them are currently dependent on government subsidies although land registries in general are able to recover their costs when land markets begin to operate. This point has not yet been reached in the countries of east and central Europe where the infrastructure for the land market to operate is being put in place but the market has not yet developed. Only in Hungary and to a lesser extent in the urban areas of Poland has the market begun to take off (Baldwin et al. 1998).

10.3 Governance in Less Developed Countries

The 'reinventing' of government and the deployment of strategies in support of more effective public administration is of particular relevance to less developed countries. What has become known as 'good governance' is increasingly being identified as a crucial determinant of their economic performance. In order to achieve good governance there must often be re-engineering.

In a report on sub-Saharan Africa, the World Bank (1989) addressed the issues of governance with unprecedented frankness. Governance was seen as encompassing a wide range of concerns such as the effectiveness of a State's institutional arrangements and decision-making processes; its policy formulation, implementation capacity, and information flows; and the nature of the relationship between the rulers and the ruled. The World Bank's report recognized two distinct but intimately linked dimensions of governance: one being political (related to the degree of genuine commitment to the achievement of good governance and the need to arbitrate equitably among competing interests), and the other technical (related to issues of efficiency and public management).

The upsurge of interest in governance reflects perceptions in which the success of market economies is being contrasted with the failure of centralized planning. The inefficiencies of state enterprises and public agencies, especially at a time of fiscal crisis, have also prompted a re-examination of the role of the State. In addition, popular discontent at the abuses of authoritarian regimes in a number of jurisdictions has spurred a search for more democratic and responsible forms of government, a process that is heightened by concerns that widespread corruption is siphoning away both domestic and foreign resources. Privatization and economic liberalization significantly reduce the power of state bureaucrats, as do policies of decentralization that 'empower' local communities. Empowerment means giving people more opportunities to decide on matters that affect them and allows them to assume greater responsibility for the delivery of public services.

Although there are many different views as to what constitutes good governance and effective public administration within a developing country context, there appears to be a number of characteristics on which there is general agreement. These include:

- political accountability;
- freedom of association and participation;
- sound judicial system;
- freedom of information and expression;
- bureaucratic accountability;
- potential for capacity building.

The first four characteristics are of a political and legal nature, beyond the scope of a discussion about the role of re-engineering land administration systems. Measures to achieve political and legal accountability need to be accompanied by additional arrangements to make bureaucracies more accountable. This requires monitoring the performance of public agencies and officials within an effective and politically autonomous system that is dedicated to correcting bureaucratic abuses and inefficiencies. Good governance requires not only political commitment, systems for ensuring accountability, sound administration of justice and freedom of information, but also competent and effective public agencies.

In regard to the privatization of public enterprises, less developed countries have distinct challenges, especially with respect to the issues of regulation and competition. A regulatory environment must be in place to prevent privatized concerns from exploiting their monopolistic position. Indeed, if the full efficiency gains that are potentially available through privatization are to be realized, it is essential that the benefits of the market place be understood. A company's relationship with government after it has been privatized must be well defined in a clear and binding relationship so that investors can predict with some degree of certainty how the company will perform.

In the end, the most fundamental issues of governance, public administration and the privatization of public enterprises in developing countries will involve an understanding of who are the ultimate clients, what are their needs, and how best to satisfy their needs. Part of the reason why, in many countries, informal settlements have sprung up is that this understanding is limited. As long as the majority of the population remain outside the formal sector and remain invisible to the policy-makers, bureaucrats, and managers of newly privatized agencies, then no fundamental reforms are possible. The benefits of reinventing government, more effective governance, re-engineering government programmes, and privatization will of necessity be felt almost exclusively by those with whom there is a formally defined relationship.

One of the special strengths of those who have recently been promoting property formalization and registration in less developed countries has been their ability to articulate effectively the importance of these programmes in the language and priorities of senior policy makers (de Soto 1993). For a government programme to meet more effectively the needs of its 'customers' it must know who those customers are and it must develop a formal relationship with them. All too often, vast numbers of people are excluded.

10.4 Organizational Structures for Land Administration

There are a variety of ways of organizing the administration of land and establishing the status of the offices responsible for the implementation and maintenance of the land and property registers. In Europe, for example, registration offices are in general under central or federal government control. Throughout most of North America, land registration is a state or provincial responsibility and day-to-day operations are often assigned to local government. As the re-engineering agenda is advanced and new opportunities are identified, a variety of organizational issues still need to be tackled, including:

- the integration of operations and the selection of a lead agency;
- the degree of centralization and decentralization;
- outsourcing and partnerships between the public and private sectors; and
- structuring land administration along business lines.

Integration and the Selection of a Lead Agency

The provision of a broader based land administration service usually runs counter to the traditional separation of responsibilities between the land registry, the state valuation service, and the national mapping or cadastral agency. The selection of a lead agency for an integrated organizational structure that will set and monitor standards and take care of national interests is always controversial. Even within government, there is inevitably competition between ministries and departments vying for power, influence, and resources.

Any proposal that a government department might lose its identity or become subservient to another agency or authority inevitably creates political difficulties. Even the idea that duplication can be reduced through standardization and the sharing of common data will result in some people feeling threatened. Similarly there will normally be strong opposition from any government department that is required to restructure its data to be in conformity with an overall national corporate database. Even an apparently small technical matter such as the adoption of a common address system can create difficulties, as it will be seen to result in extra work, more expense, and possibly less internal efficiency for the organization concerned.

If government is to treat all its data (other than those classified for security reasons) as a corporate resource, then there must be some common standards to ensure compatability between systems. Even if a committee creates these standards, there must be leadership from someone or somewhere. Since in many countries the land registry function may fall under the Ministry of Justice, the valuation under the Ministry of Finance, and the mapping and cadastre possibly under the Ministry for Physical Planning, the leadership may need to come from above the Ministers concerned. As a result, responsibility will then lie at Cabinet level or in the Prime Minister's Office.

In some countries in east and central Europe, for instance in Latvia and in Slovenia, the cadastre, valuation, and physical planning all fall under the same ministerial portfolio although the administration of the Land Books is basically the responsibility of the Ministry of Justice. In the State of Victoria in Australia the offices of the Registrar of Titles, the Surveyor General, and the Valuer General have all been combined within one Ministry (a resurrection of the old lands department concept). In Austria the need for data integration led to close cooperation between the cadastral authority that handles the mapping and the Ministry of Justice that is responsible for the Grundbuch; together they created a joint land and property database. In Sweden, re-engineering led to the creation of a new authority that oversees both the real estate database and the National Land Survey. What began as a series of projects to computerize parts of the land administration process has ended up with fundamental institutional reform.

Integration can come through the leadership of a strong and persuasive individual. In the early days of the development of land information services it was almost always an individual who was the catalyst for change and encouraged agencies to take advantage of the new technologies. Behind every implementation of an integrated system there have inevitably been one or two people responsible for its success or failure. Systems were built bottom up and outwards around the ideas of a few articulate visionaries.

Now that the role of information and the opportunities created by the technology are better understood, a more institution-driven approach is emerging. Throughout much of east and central Europe, for example, there has been active debate about the role of the cadastre in the new information age. In spite of this, no standard model has emerged and each country is adopting its own solution to the problems of coordination and the selection of a lead agent.

A number of strategies can be adopted to increase cooperation between agencies and hence give rise to greater efficiency. These include:

1. simple sharing of physical facilities by different property agencies;
2. sharing of staff, for example, those who work on a front desk receiving documents and fees;
3. sharing of data with individual agencies being given primary or 'custo dial' responsibility for selected data—for example the land registry might be responsible for the parcel identifier, owner's name, and current legal

status, while the property assessment agency may be responsible for information about improvements to the property and its asset;
4. sharing of common functions (such as the use of common forms) and procedures for similar functions; and
5. sharing of a common database environment, for example, by using standardized inquiry and communications software to provide access to a family of databases.

One of the more successful examples of integration is in the Canadian province of New Brunswick where the provincial land registry, surveying and mapping, personal property registry, property valuation, and land information coordination functions have been made the responsibility of one agency, Service New Brunswick. Previously, each one of these functions had people and facilities assigned exclusively to it. The agency now shares these resources throughout the province, has developed common standards for data, and is currently constructing an integrated database environment (using Internet and Intranet technologies to provide on-line access).

Centralization and Decentralization

Detailed land administration operations may be centralized or decentralized depending on such matters as the size and political geography of the country and the nature of its transportation and communications. Centralization can lead to economies through the use of common administrative procedures, standardization in documentation and the easier exchange of information between users, and economies of scale in which large and powerful systems can be used with mass production techniques. Decentralization of most routine operations on the other hand can offer important advantages, especially in countries where distances are large or travel is inconvenient. From a political perspective, bringing government closer to the people through decentralization has considerable appeal (UNECE 1996).

From a practical point of view, the location of land administration offices at the district or local government level helps to ensure that systems are more answerable to the local community. The local community is often reluctant to travel long distances to visit land offices and as a result, especially in less developed countries, transfers are likely to take place without being recorded in the formal administrative system. The existence of local offices is likely to result in more reliable and up-to-date records through greater community involvement. Conversely, administrators who are close to the people and their land know and understand their needs.

In many countries there are examples where inheritance has not been registered or where dealings that are not officially recognized have informally been carried out—for instance the leasing of land in areas where bureaucratic controls make formalization difficult or where the State does not officially recognize leasehold tenure. Decentralization should allow the overall land administration process to operate more efficiently and effectively and permit the system to respond to local community needs.

Where decentralization takes place there must generally be strong central control and good communication between the local offices and their headquarters. In Latvia, for example throughout the land restitution process, each Land Book Judge tended to operate independently, resulting in different decisions being taken depending on the location of the land and the experience of the Judge. Latvia has a population of four and a half million people (of whom one-third live in and around the capital city, Riga) and has twenty-eight Land Book Offices. In larger countries the problems are more acute—in Thailand for example there are approximately 500 offices with 13,000 staff while Indonesia has 300 offices with 26,000 employees. Such cases have additional re-engineering, communication, and human resource management challenges (Holstein 1996).

Outsourcing and Partnerships between the Public and Private Sectors

Although the ultimate responsibility for any land administration system lies with government, in many countries the private sector plays a significant role in the day-to-day functioning of the system. The technical processes of surveying and mapping and even valuation may be subcontracted to the private sector. Thus private surveyors may undertake control surveys, topographic base mapping, the detailed measurement and recording of property boundaries, or the valuation of properties in accordance with agreed criteria. In St Lucia, the initial creation of the registers was carried out by the private sector; this was somewhat unusual since first registration is normally the responsibility of government officials. Conversely, it is usual for much of the maintenance of the system to be done by private surveyors who undertake subdivision surveys or the re-establishment of old boundaries while private lawyers are often involved in the land transfer process.

In many countries there is insufficient capacity in the public sector for the government to undertake all aspects of the creation and maintenance of the land administration function. Furthermore it is increasingly the official policy to encourage greater private sector involvement and there is a general trend towards the 'outsourcing' of work that has traditionally been regarded as the sole responsibility of government staff. As noted previously, the reasons for this are in part the recognition that the private sector can do certain tasks more cheaply and more quickly, and partly because of a wish to decrease the number of civil servants by 'downsizing' government departments.

Land administration data must be subject to quality control, whether these data are gathered by the private sector or by the government. Where private surveyors are employed to measure property boundaries there is often a system of licensing but this does not in general apply to valuers. While it is common to license individual surveyors, the licensing of companies is unusual although the International Standards Organization (ISO) certification is becoming more widespread. Where a licensing system exists, checks may still be applied to the work of the private surveyor. Depending on the manner in which these checks are undertaken, quality control may be very expensive especially if attempts are made to check all aspects of each cadastral job carried out by the private sector. Sample checks and a sensible understanding of risk management can overcome this difficulty. In such cases it is helpful if there is a strong professional body that enforces discipline on its members and requires its members to carry professional indemnity assurance so that in the case of error clients can receive compensation.

The balance of responsibility between the public and private sectors depends ultimately on:

1. each country's political objectives with regard to privatization;
2. the distinction between juridical and technical work;
3. the nature and traditions of the particular jurisdiction;
4. available funding;
5. questions of access to certain types of data and the need for privacy; and
6. the strength of the private sector (UNECE 1996).

Structuring Land Administration along Business Lines

In recent years, attention has been given to streamlining the costs of land administration, enhancing its revenue capabilities, and requiring a more businesslike approach. Three models have emerged to date:

1. Reformed state agencies. These operate within a ministerial environment, but their chief executives are charged with meeting established targets for revenues and expenditures and they are held accountable for the outcome. Sometimes referred to as operating under a Trading Fund, examples may be found in the UK and in the New South Wales Land Titles Office.
2. Government corporations. These have more independence (with their own board of directors and multi-year business plans) although they are ultimately responsible to their shareholder, the government. Service New Brunswick, which we have discussed in some detail, fits this model inasmuch as it is financially self-sufficient and depends entirely upon fees derived from services and the provision of information.
3. Private utilities. These may consist of a private company or a public-

private sector partnership. In either case, private capital is invested in improving the registry function and it is then run as a private business for some agreed-upon period of time (for example, ten years). An example is Teranet in Ontario, Canada, which functions as a regulated utility in the sense that the government retains some control over the structuring of fees and has ultimate public custodial responsibility for the registered documents.

These models have been applied to well-established land administration systems where there is access to investment funds, sufficient land market activity to allow agencies to operate without subsidies, and a political will to grant them a degree of independence. Among the poorer nations of the world such innovation has not occurred although some moves towards greater autonomy have begun.

References

Baldwin, R. A., Bogaerts, T., Csaki, C., Dale, P. F., and Zichy, A. (1998), 'The Development of Land Markets in Central and Eastern Europe', Report to the European Union Action for Cooperation in the Field of Economics.

Butler, R. (1994), 'Reinventing British Government', *Public Administration*, 72.

De Soto, H. (1993), 'The Missing Ingredient', *The Economist*, 11 Sept.

Holstein, L. (1996), *A Review of Some World Bank Experience with Land Titling and Registration Projects* (Washington: World Bank).

Rhind, D. (ed.) (1997), *Framework for the World* (Cambridge: Geoinformation International).

United Kingdom Government (1996), *Government. Direct: A Prospectus for the Electronic Delivery of Government Services* (London: Stationery Office).

United Nations Economic Commission for Europe (UNECE) (1996), *Land Administration Guidelines: With Special Reference to Countries in Transition* (New York and Geneva: United Nations).

World Bank (1989), *From Crisis to Sustainable Growth: Report on Sub-Saharan Africa* (Washington: World Bank).

11

Economic Issues in Land Administration

11.1 Basic Issues

Land is of such fundamental importance that the land administration function has tended to be taken for granted. Increasingly, however, there is a debate as to how much money should be allocated to this area and with what priority. A host of concerns have been raised with respect to:

1. documenting the benefits and costs of titling and registration projects;
2. financing the construction and ongoing management of land administration infrastructure;
3. developing appropriate pricing strategies and policies for land administration services and products; and
4. examining the economic issues associated with determining the most effective roles for government and the private sector in the land administration field.

Where more fundamental assessment of the role of real property has taken place, two schools of thought have emerged that are not mutually exclusive. The first has been based on traditional arguments for detailed a priori benefit/cost assessments (factoring in both quantifiable and non-quantifiable variables); the second and more recent has argued for minimal initial investment in the infrastructure, leaving it to market forces to dictate subsequent developments.

11.2 Benefits of Titling and Registration

The classic work of Gershon Feder and his World Bank colleagues on assessing the benefits of titling and registration has recently been reported in Feder and Nishio (1998). Feder developed a conceptual framework for the economics of land registration, initially in the context of a study on rural Thailand (Feder et al. 1988). Two links between titles and economic performance were highlighted: the enhancement of tenure security and the role of titles in collateral arrangements that would facilitate access to institutional credit.

Feder's conceptual framework for evaluating landownership security and farm productivity is illustrated in Figure 11.1.

Using empirical evidence from rural Thailand, Feder and his team compared the economic performance of two groups of farmers: one group was without legal titles and operated in forest reserves while the another group had legal titles and operated outside the forest reserve boundaries. Study sites were selected from four provinces, with the comparative groups operating in

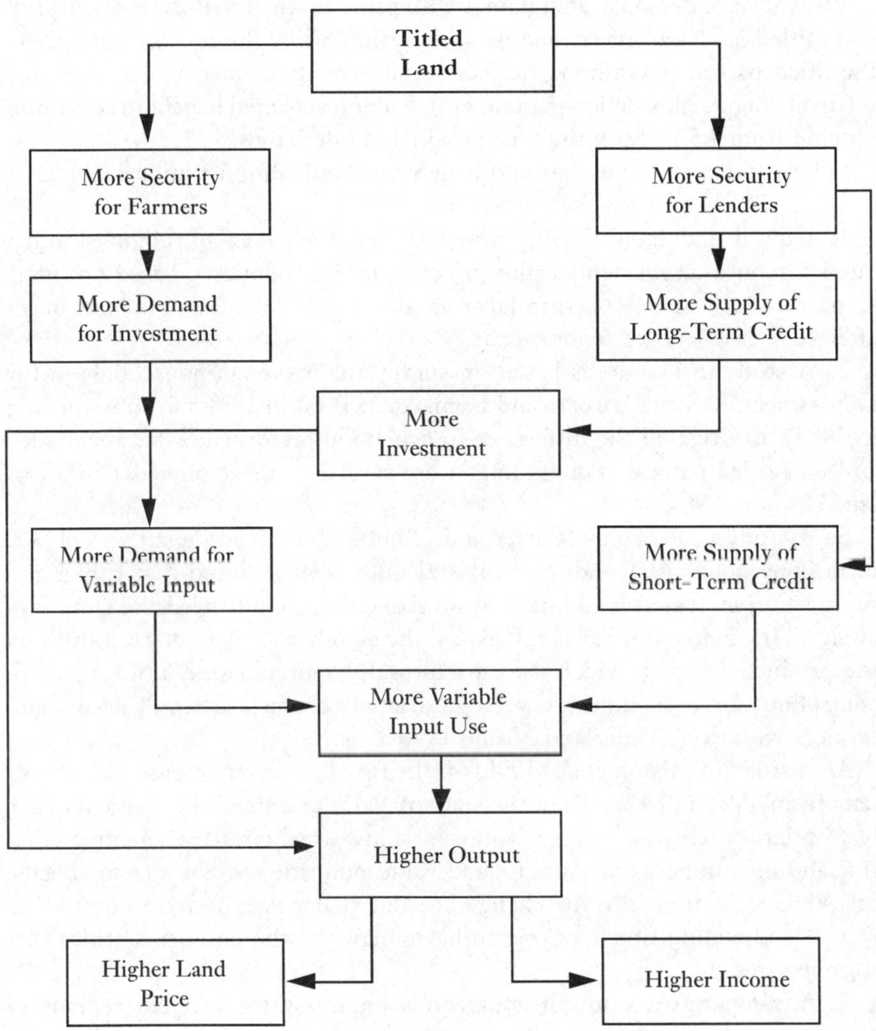

FIG. 11.1 Ownership security and farm productivity

Source: Feder et al. 1988

geographical proximity and within a similar agrarian and climatic environment. The effect of land registration on farmers' access to credit was clearly established, as were differences in economic performance between titled and untitled farmers (Feder et al. 1988). However, in order for the economic analysis to be relevant for policy discussions, 'the benefits of land registration needed to be discussed in terms of social benefits, instead of benefits to individual farmers, and in comparison with the costs of land registration to be borne by the society' (Feder and Nishio 1996).

Feder's work demonstrated (albeit with some caveats) that the market price of untitled land was lower than its social value, while the reverse was true of the titled parcels. Examining the net social benefits of ownership (assuming relatively high-risk aversion parameters), Feder discovered benefit to cost ratios ranging from 4.5 to 12 in the four provinces under study.

Other studies into the costs and benefits of land administration include:

1. Dowall and Leaf (1990), who analysed the effect of, amongst other factors, tenure on residential plot prices in urban Indonesia, based on interviews with real estate brokers in Jakarta. Land prices were found to be strongly affected by the level of tenure security.
2. A study in Honduras based on surveys of 450 farm households in the Department of Santa Barbara and Comayogua (first in 1984 and subsequently in 1994) investigated the difference in economic performance between titled and non-titled parcels. The findings were similar to those of Feder in Thailand (Lopez 1996).
3. A study in Paraguay (Carter and Olinto 1996) based on surveys of 300 farm households was conducted in 1991 and 1994. It found that titling and registration increased land-attached investments and enhanced the supply of credit. The main empirical finding was 'the significant effect of land titles on the productivity with which the agricultural resource base is utilized, when controlling for both the observable and latent characteristics of farms and farmers with title' (Feder and Nishio 1996).
4. A study by Alston et al. (1996) on the Brazilian frontier, based on survey data from 1992 and 1993 from the state of Para, examined the characteristics of frontier settlers, land tenure, land values, investments, etc. It concluded that title and investment contribute to land value, and title promotes farm-specific investment. Additionally, the change in value that was expected to occur as a result of obtaining title appeared to have increased the number of titles that were registered.
5. A comparative study of indigenous tenure systems in ten regions of Ghana, Rwanda, and Kenya was based on farm surveys conducted during 1987–8. Contrary to experiences in Asia and South America, this found little

link between land registration and the economic benefits (Migot-Adholla et al. 1991). Migot-Adholla concluded by arguing that land registration is 'unlikely to be economically worthwhile for much of sub-Saharan Africa at this stage of economic development' except in cases where there are many land disputes, where the indigenous systems are particularly weak, or where major development projects are planned.

These studies have focused on the benefits of titling per se without the added value that comes from operating an integrated set of land policies and a national land information service.

11.3 Lessons Learnt

In reviewing the various studies described above, Feder and Nishio addressed two broad policy questions:

1. What are the key prerequisites for economic viability of titling and registration programmes?
2. Where registration is appropriate, what are the key factors in ensuring that these programmes are conducive to achieving equitable social outcomes?

They established a number of principles for effective and equitable titling and registration, including the need for:

1. reasonably well-functioning financial markets that can extend long-term credits when land is used as collateral;
2. clear incentives to increase output per unit area, through investment in land-attached investments and use of more variable inputs, in terms of existing economic opportunities;
3. where incentives for productive investment do exist, the necessity to ascertain whether the existing tenure system provides sufficient security (measured, for example, by the prevalence of land disputes);
4. when there is latent demand constrained by lack of secure title, the necessity to assess the extent to which additional regulatory restrictions affect land transactions; for example, complicated land use controls may diminish the benefits of registration by increasing transaction costs and limiting land market activities.

They also recognized that titling and registration projects have social as well as economic impacts, and in particular that:

1. investments in titling and registration, if not properly implemented, can produce undesirable and unintended outcomes;

2. titling may provide opportunities for 'land grabbing' by those better informed, more familiar with formal processes, with better access to officials and the financial means to register documents; and
3. there is the potential for wealthy individuals having advance knowledge to acquire untitled land at low cost.

The principles they advocated in order to reduce the risks of undesirable social impacts included:

- transparency: a land registry should be administered in a way such that the administrative procedures and fees for titling and registration are clear to all participants.
- consultation: landholders should be closely involved in the processes so that they can take ownership of the system. 'Consultation with the beneficiaries, reliance on communities and community-based organizations, simplicity of procedures and speed of delivery can be used to enhance landholder involvement and improve the prospects for equitable outcomes' (Feder and Nishio 1996).
- cost effectiveness: a major factor in designing systems and processes should be the focus on costs with an emphasis on maximizing net social benefits and improving sustainability (systematic rather than sporadic registration in Thailand, for example, has contributed to keeping down the costs).
- cost minimization: costs should be kept low for first-time registrations so as to provide incentives for low-income landholders to register their land.
- customer awareness: there is a need always to keep in mind that titling and registration programmes are created to serve the people.

While there is substantial support for these principles, there is increasing criticism of the economic arguments that underpin them on the grounds that:

1. many of the studies have been too limited and too superficial;
2. apart from access to credit, the relationship of property to a large number of economic and social activities has been ignored;
3. there has been limited understanding of the technical and resource challenges and opportunities involved in sustaining the infrastructure over time; and
4. the work has been too much oriented towards state planning and priorities, ignoring much of the potential role of the market in determining what should be done.

The consensus view is that economic analysis is still at a very early stage of development. To date it has been easier to criticize what has been achieved than to propose better models and more effective tools of analysis based on scientific principles.

11.4 Financing Land Administration Infrastructure

The initial financing of land administration infrastructure (registry systems, surveying and mapping programmes, etc.) has traditionally come from government sources; increasingly the maintenance and update of systems is being funded from the revenues generated, for instance through the selling of information or the provision of services. This is creating particular difficulties for the mapping agencies whose ability to survive on the basis of sales of maps and surveying services is limited. National mapping agencies are also exposed to the danger that, if commercially successful, they may then be sold off to the private sector.

In contrast, land registries have in general had a monopoly with little pressure from government to open this to competition. In several countries (Sweden, Great Britain, etc.) the land registry system is expected to operate at a marginal but not excessive profit. According to *The Economist* (1998) the overall transaction costs including survey, legal, and registration fees but excluding taxes on a property worth about US$200,000 ranges from about 3 per cent in Australia and in England, through 5 per cent in Sweden, 6 per cent in the United States, 8 per cent in South Africa, to 10 per cent in Portugal.

There are at least three different ways of financing land administration operations, namely from taxes, by fees, and by commission. Financing out of taxation means that there is no direct connection between the activity from which the tax is raised and the grant that is given by the government; there is thus no direct incentive to optimize the service delivered. Financing by fees means that to some extent the user pays for the service; tariffs may be decided by the government and some or all of the fees may go directly or indirectly to the agency. Financing by commission means that an applicant pays for the service and that the agency has the authority to decide about the tariff based on rules set by the government. A mixture of approaches may be used within any given jurisdiction. In Sweden, for example, activities where there is a direct applicant are generally subject to a fee or commission, while activities that are more connected to the rule of law or the overall public good have been financed by grants.

The optimum approach will depend on government policies and in particular on attitudes towards:

1. the extent to which the organization should be allowed to influence its own income through active marketing of its services and be allowed to expand accordingly;

2. the extent to which the organization should be allowed to borrow money in the regular market for investments;
3. the manner in which the organization should implement an independent system of accounting;
4. whether the organization should be allowed to fix its own level of staff remuneration;
5. whether the organization should be authorized to decide on its own internal organization in divisions etc., and where it should establish offices; and
6. whether the organization should have an independent board of directors, or be controlled directly from the political level (UNECE 1996: 52–3).

In many developing countries, there is a need to restructure the fundamental organization and legislation for cadastral activities, land registration, and for the national information infrastructure so that they become on a par with more developed market economies. Invariably, this requires funding that is not available from internal sources. Traditional recourse has been to seek support from the various development banks and aid agencies for grants or loans. In some countries with high near-term growth prospects and relatively low 'country risks' there may be an opportunity to turn to the private financial markets but in poorer nations this is not yet the case. In these countries external funding is the only option.

While the initial creation or re-engineering of land administration systems may require subsidies, there is in many jurisdictions increasing pressure to fund some or most of the ongoing operations through services sold to the public. This is the case in both developed and developing jurisdictions. It has also led to the emergence of a variety of pricing strategies especially with regard to charging for on-line services—a topic receiving increasing attention in the literature under the heading of 'electronic commerce'.

References

Alston, L., Libecap, G., and Schneider, R. (1996), 'The Determinants and Impact of Property Rights: Land Titles on the Brazilian Frontier', *Journal of Law, Economics and Organization*, 12/1.

Carter, M., and Olinto, P. (1996), *Decomposition and Identification of the Impact of Land Titling on Agricultural Productivity in Paraguay* (Madison: Department of Agricultural Economics, University of Wisconsin).

Dowall, D., and Leaf, M. (1990), *The Price of Land for Housing in Jakarta: An Analysis of the Effects of Location, Urban Infrastructure and Tenure on Residential Plot Prices* (Berkeley: report prepared for the Regional Housing and Urban Development Office, USAID).

Economist (1998), 'Buyers' blues', *The Economist*, 5 Sept.

Feder, G., and Nishio, A. (1996), 'The Benefits of Land Registration and Titling: Economic and Social Perspectives', Orlando, Florida: paper presented to the International Conference on Land Tenure and Administration, November.

————(1998), 'The Benefits of Land Registration and Titling: Economic and Social Perspectives', *Land Use Policy*, 15/1.

——Onchan, T., Chalamwong, Y., and Hongladarom, C. (1988), *Land Policies and Farm Productivity in Thailand*, (Baltimore: Johns Hopkins University Press).

Lopez, R. (1996), *Land Titles and Farm Productivity in Honduras* (College Park, Md.: Department of Agricultural and Resource Economics, University of Maryland).

Migot-Adholla, S., Hazell, P., Blarel, B., and Place, F. (1991), 'Indigenous Land Rights Systems in Sub-Saharan Africa: A Constraint on Productivity', *World Bank Economic Review*, 5/1.

United Nations Economic Commission for Europe (UNECE) (1996), *Land Administration Guidelines* (New York: United Nations).

12

Human Resources Management

12.1 The Importance of Human Resources Management

Effective human resources management is a key ingredient in building and sustaining a country's land administration infrastructure. Whether it is building new systems or reforming existing ones, the recruitment, training, provision of support for, and evaluation of employees will ultimately be far more important than matters pertaining to technology and process. Yet traditionally, human resources management has not been given much serious attention in the land administration field.

Within the broader public administration arena, however, the human resources management function is increasingly being recognized as a central organizational concern and that 'its performance and delivery are integrated into line management; the aims shift from merely securing compliance to the more ambitious one of winning commitment. The employee resource, therefore, becomes worth investing in, and training and development thus assume a higher profile' (Storey 1991).

What distinguishes modern human resources management from the more traditional personnel functions is its focus on utilizing human resources to strategic management objectives. Effective human resources management seeks to:

1. link human resources management issues to the overall strategy of an organization;
2. build strong organizational cultures aimed at uniting employees through a shared set of goals and values ('quality', 'service', 'innovation', etc.) and by promoting a commonality of interests amongst employees and management;
3. recognize employees as a resource, as social capital that can be developed and can contribute to competitive advantage;
4. replace traditional top-down communication, coupled with controlled information flow, to a sharing of information and knowledge; and
5. achieve flexibility and adaptability to manage change and innovation in response to rapid changing circumstances (Burt and Spector 1985).

This section examines briefly some of the principal human resources management issues, particularly as they relate to developing countries. The focus will be on concerns within the public sector (where most of the core land administration activity occurs), the broader issues of developing local capacity in both the public and private sectors, and the requirements for developing professional associations.

12.2 Public Sector Reforms

Significant emphasis has been given in recent years to the challenges of building and sustaining institutions for capable public sector administration in the developing world. This has arisen from an increased awareness that public sector bureaucracies 'need to be fundamentally altered if countries are effectively to introduce market economies, sustain democratic political systems, and address pressing social needs and new international conditions' (Grindle 1997).

Unfortunately, in many developing countries, the state sector, far from providing the institutional foundation for economic and social development, has become an oversized and inefficient competitor for scarce resources and a major source of corruption. The need for public sector reform featured prominently, for example, in the World Bank's 1997 World Development Report. The Report described in detail a series of institutional building blocks that could support the evolution of an effective public sector and discussed a series of promising options for their implementation. In describing the foundations for an effective public sector, the Bank noted that this 'focus on institutions is very different from the traditional approach of technical assistance, which emphasizes equipment and skills and administrative or technical capacity. The emphasis here is on the incentive framework guiding behaviour—what government agencies and officials do and how they perform' (World Bank 1997).

In a 1994 study on public sector capacity sponsored by the United Nations Development Programme, six countries (Bolivia, Central African Republic, Ghana, Morocco, Tanzania, and Sri Lanka) were chosen for an examination of good and poor performance. The characteristics that appeared to differentiate good performance included 'strong mission focus, performance-oriented management practices, high performance expectations, and organizational autonomy in personnel matters [which] tended to go together in ways that allowed some organizations to develop and inculcate a set of norms for organizational performance and individual and group behaviour that consistently affected their ability to perform responsibilities more effectively than others' (Grindle 1997).

Building on this and other studies, the World Bank Report identified three essential building blocks for building an effective public sector: strong central capacity for formulating and coordinating policy; efficient and effective delivery systems for public services; and motivated and capable staff. The third block is of particular relevance to this discussion since 'whether making policy, delivering services or administering contracts, capable, and motivated staff are the life-blood of an effective state' (World Bank 1997).

Some of the key elements required to build a motivated and capable staff include merit based recruitment and promotion, fostering a sense of *esprit de corps*, and providing adequate compensation. The need for effective performance pay systems is a matter of special concern.

12.3 Effective Performance Pay Systems

Pay is one of the most important and contentious elements in the employment relationship. Increasingly it is recognized that some kind of flexible pay system, closely tied to performance criteria, is essential for attracting and retaining skilled staff. Recent attention has focused on skill-based pay systems in which pay increases are linked to the skills an employee acquires and applies. Such increases are generally in addition to, and not instead of, any general pay increases that employees may receive (de Silva 1998).

In such systems, pay increases are usually tied to three types of skills:

1. horizontal skills that involve a broadening of skills in terms of the range of tasks;
2. vertical skills that involve acquiring skills of a higher level; and
3. depth skills that involve a high level of skills in specialized areas relating to the same job.

Skill-based pay is person based, not job based, rewarding a person for what he or she is worth rather than what the job is worth. Where effectively implemented, it promotes the development of a flexible and multi-skilled workforce, enhances productivity and quality through better use of human resources, and provides incentives for self-development. At the same time there are several problems that generally need to be addressed. These include the need to have effective training programmes; strategies for dealing with obsolete skills; the need to clearly describe expectations and requirements; the fact that some skills may be paid for but used infrequently; and difficulties in anticipating future skills needs.

The International Labour Organization has published guidelines for introducing performance and skill-based pay systems (de Silva 1998). These guidelines recognize that some of the key factors that can contribute to the success of skill-based pay include the employer's commitment to continuous training and development; dismantling strong bureaucratic hierarchies and narrow job demarcations; and the introduction of participative management practices.

12.4 Controlling Corruption

Corruption poses its own special challenges for public sector reform, and it is an area where land administration agencies are not immune (especially in their regulatory and taxation functions). As the World Bank reported:

> Corruption flourishes where distortions in the policy and regulatory regime provide scope for it and where institutions of restraint are weak. The problem of corruption lies at the intersection of the public and private sectors. It is a two-way street. Private interests, domestic and external, wield their influence through illegal means to take advantage of opportunities for corruption and rent seeking, and public institutions succumb to these and other sources of corruption in the absence of credible restraints (World Bank 1997).

The causes of corruption are many, but the incentives for corrupt behaviour by public officials arise in an environment of wide discretion and limited accountability. Corruption is especially endemic in environments where there is considerable 'red tape', where there are few audit and review mechanisms, where there is limited oversight in the awarding of contracts, and where salaries are very low.

Corruption cannot be tackled effectively in isolation from other problems. It is a symptom more than a cause of problems and needs to be controlled through a variety of strategies, including developing merit-based recruitment and promotion systems, and performance-based pay systems as discussed above. There will also be the need to reduce the opportunities for corrupt behaviour (such as by limiting the discretionary authority of officials) and to strengthen accountability through various mechanisms of monitoring and punishment.

While employers, both in the public and private sector, have key roles to play in fighting corruption, so do professional associations. In the words of the International Federation of Surveyors:

> A professional is distinguished by certain characteristics including:
>
> - mastery of a particular intellectual skill, acquired by education and training;
> - acceptance of duties to society in addition to duties to clients and employers;
> - an outlook that is essentially objective; and
> - the rendering of personal service to a high standard of conduct and performance.

Professional surveyors are increasingly being taught that their ethical responsibilities extend to the public, to their clients and employers, to their peers and to their employees. Accordingly they acknowledge the need for integrity, independence, care and competence, and a sense of duty (FIG 1998*a*).

Many professional bodies have established similar codes of ethics and business conduct in order to encourage their members to show care, a sense of duty, and a concern not only for their clients and fellow professionals but also for the community at large. In 1998, the Royal Institution of Chartered Surveyors (RICS 1998) revised its rules of conduct and disciplinary procedures, reinforcing its standards of business and professional conduct in the light of changing market practices. The rules deal not only with such matters as the keeping of accounts and the handling of clients' money but also conflicts of interest, impartiality, and independence. With the deregulation of many forms of service delivery in which more and more is put under the control of the market rather than direct government, there is an increasing need for self-discipline. This is one role for professional associations.

12.5 Building Local Capacity

Both the public and private sectors in developing countries have had to address the special challenges of identifying, recruiting, and retaining people with specialized expertise. In the land administration field, this is particularly true in building the necessary cadre of senior lawyers, engineers, and computer specialists. Also, the increasing influx of women into the workforce has raised special issues relating to gender discrimination, opportunities for training, and promotion and welfare facilities.

Increasingly, the international lending agencies are recognizing the importance of fostering local capacity development, with the 'traditionally dominant donor role in decision-making and management giving way to recipient ownership of the development process' (Michel 1995). This entails, among other things, making sure that local officials are responsible for the overall management of development programmes, and that locally available expertise is used optimally. The role of foreign experts provided under technical assistance programmes is tailored to specific, short-term requirements with local counterparts being trained to take over at the end of programme.

The Organization for Economic Development and Cooperation (OECD) has recommended that aid agencies can assist local capacity building by:

1. giving more recognition in aid programmes to strengthening local capacities and giving the use of local expertise priority;
2. providing more time and longer-term funding for adequate participatory frameworks to be established;

3. facilitating more cooperation and exchange of experience between developing countries themselves, especially drawing on regional expertise; and
4. strengthening institutional pluralism in civil society through, for example, supporting professional associations, academics, advocacy groups, the media, entrepreneurs, trade unions, etc. (Michel 1995).

12.6 Education and Training

Education and training present huge issues to the land administration field. Traditionally, the professional staff have come from a wide variety of disciplines (law, surveying, engineering, etc.). Their education has tended to be relatively narrow, providing only limited understanding of the role of property in economic and social development, and the challenges associated with building and sustaining effective land administration infrastructures.

This is beginning to change as an international network of universities with special expertise in land administration is emerging and as a number of professional societies start to address the field more holistically. A multidisciplinary approach is necessary if the complex problems of land and property development are to be addressed in a way that ensures long-term sustainability. Modular structures in education permit a much more multidisciplinary approach while within the professions there is a greater awareness of the need to work with and especially understand the language of other professionals.

Beyond the domain-specific education and training requirements of the land administration field, there is a growing awareness that employees need not only specialized technical skills, but also the skill of learning how to cope with continuous and radical changes. Demands are emerging for new forms of training that are flexible, available on demand, and interactive.

In particular, there are seven groups of skills that are increasingly desired by employers, namely:

1. Knowing how to learn—especially the capacity to collect, analyse, organize, and apply information. This covers techniques, attitudes, and knowledge that both facilitate processing of information and enable workers to adapt quickly to new demands.
2. Reading, writing, and computation skills; these are required in the workplace and involve increasing interaction with sophisticated technology.
3. Communication skills—especially the ability to speak and listen effectively in an environment where more interaction with other sectors and customers is being demanded.
4. Adaptability skills—the ability to solve problems and think creatively in an environment of increasingly decentralized decision-making.

5. Developmental skills—the ability to manage personal and professional growth based on self-esteem, motivation, goal setting, and self-discipline.
6. Interpersonal skills, teamwork and negotiation skills—to ensure that each individual can work in a group, has the ability to judge and balance appropriate behaviour, absorb stress, deal with ambiguity, share responsibility, inspire confidence, and recognize and cope with various personalities.
7. Leadership skills (Tan 1997).

All of the above may be subsumed under the name of management skills. University education has in the past focused very much on single academic disciplines (for instance the study of a language, a science, or an art) and many curricula have made no mention of management skills that have been regarded as not being academic. Specialist management colleges have filled the void. More recently the practical based academic subjects such as engineering and surveying have introduced more elements of management training into their courses.

Important elements in developing training programmes include:

1. relating the training goals to the strategic needs of the organization;
2. providing all levels of employees with opportunities for lifelong learning;
3. requiring workers to be accountable for learning new skills (developed around individual training and development plans for each individual);
4. selecting recruits more carefully so that training can have the greatest impact on productivity; and
5. designing a training curriculum that stresses corporate citizenship, a contextual framework, and has a core on which to focus (Tan 1997).

12.7 The Role of Professional Associations

Given that there will be a growth in private sector activity within land administration, the quality control of the work of private practitioners takes on greater importance. One answer has been the licensing by government of conveyancers, valuers, and cadastral surveyors. In all the central and eastern European countries currently restructuring their land administration systems, there has been a debate on whether and how to use the private sector while maintaining the quality that government officials often claim is only possible within the government service. Most are however being forced to make use of the private sector because there is insufficient capacity within the public sector to meet the demand. Several countries have recently introduced a licensing system, preferring this to self-regulation through professional associations.

Relatively few attempts have been made to define the nature of professionalism and what constitutes a profession—the most extensive being in Carr-Saunders and Wilson (1933) and Bennion (1969). A simple definition of 'professional' is one who is paid to do a job, unlike an amateur who in theory acts without financial reward. Today the term 'amateur' is often used in a derogatory sense to imply someone who performs inadequately while 'professional' may be taken to mean competent and efficient. Many technicians are however competent and efficient and though claiming to be professional in what they do, they would not necessarily claim to be 'professionals'. From a technical perspective, the technician is often more proficient in certain aspects of his or her work than the so-called professional whose skills may lie more in the field of management. The term 'professional' then implies someone with a higher academic standard of education.

It may also imply that the person concerned is a member of a professional body that regulates his or her field of expertise. Professional institutions may operate under formal legal statute or through their own statutes and internal rules. The objectives of such bodies are to:

1. ensure that the profession is recognized and that the skills of its members are utilized wherever and whenever they are best suited to the job in hand;
2. keep actual and perceived manpower needs under review and to encourage the provision of academic programmes to train the right numbers of people to the required standard;
3. help attract the right quality of recruits onto the training programmes and into the profession;
4. set standards for and ensure that graduates from academic training programmes receive supervised practical training prior to their becoming fully qualified professionals;
5. confer professional qualifications;
6. set standards for and ensure the provision of facilities for post-qualification training;
7. set and enforce codes of conduct and standards of practice;
8. maintain a library, publish a professional journal, organize conferences and meetings and otherwise provide services to members of the profession;
9. maintain contact with professional societies in other countries and participate in the work of relevant regional and international organizations;
10. contribute to the development of policies and offer advice on those matters on which the profession is competent to express an opinion (Woolley 1991).

The International Federation of Surveyors has issued guidelines for helping to establish professional associations (FIG 1998*b*). Four reasons for forming an association are to unify the profession, to provide continuing professional development, to act on behalf of the profession especially when dealing with governments, and to contribute to society's well-being.

There has been a growing trend towards deregulation, the greater dependence on market forces, and the outsourcing of work to the private sector that was traditionally undertaken by government officials. Strengthening professional associations is one way to reduce the time and cost involved in the quality control of private sector work done on behalf of government.

References

Bennion, F. A. R. (1969), *Professional Ethics* (London: Charles Knight).

Bert, M., and Spector, B. (1985), 'Corporate-wide Transformations in Human Resource Management', in R. E. Walton and P. R. Lawrence (eds.), *Human Resource Management: Trends and Challenges* (Boston: Harvard Business School Press).

Carr-Saunders, A. M., and Wilson, P. A. (1933), *The Professions* (Oxford: Clarendon Press).

De Silva, Sriyan (1998), *An Introduction to Performance and Skill-Based Pay Systems* (Geneva: International Labour Organization).

FIG (1998*a*), *FIG Statement of Ethical Principles and Model Code of Professional Conduct*, FIG Special Publication No. 17 (London: FIG).

—(1998*b*), *Constituting Professional Associations*, FIG Special Publication No. 16 (London: FIG).

Grindle, M. (1997), 'Divergent Cultures? When Public Organizations Perform Well in Developing Countries', *World Development*, 23/4.

Michel, James (1995), *Development Co-operation* (Paris: Organization for Economic Cooperation and Development).

Royal Institution of Chartered Surveyors (RICS) (1998), *Rules of Conduct and Disciplinary Procedures* (London: RICS).

Storey, John (ed). (1991), *New Perspectives on Human Resource Management* (London: Routledge).

Tan, Peng Boo (1997), 'Human Resource Development in Asia and the Pacific in the 21st Century', Turin, Italy: paper presented at the International Labour Organization Workshop on Employers' Organizations.

Woolley, J. (1991), 'The Royal Institution of Chartered Surveyors', a descriptive leaflet written for newly industrialized countries (London: RICS).

World Bank (1997), *World Development Report 1997* (New York: Oxford University Press).

13

New Directions in Land Administration

13.1 The Changing Scene

The role of property in fostering good governance, robust economies, and strong civil societies has received fresh attention in the wake of the dramatic global changes that have occurred during the past decade. Innovative and cost effective ways to formalize property rights have emerged, linking these with new strategies and tools for building and maintaining the infrastructure necessary for sustaining a property regime. Land administration functions have been re-engineered and there have been legal reforms that have focused on modernizing, standardizing, and simplifying legislation relating to land and property. There have been new concepts of risk management, the introduction of new technologies, and a variety of organizational reforms.

Many of these reforms have been the consequence of political changes, especially as a result of the collapse of communism, the adoption of a market driven approach to the economy and the impact of information technology. The processes of re-engineering have focused on a diverse package of measures dealing with land tenure security, land and property transactions, and access to credit. They have also been concerned with the provision of support for physical planning, the sustainable management and control of land use and of natural resources, and facilitating real property taxation. Internationally funded projects have also been concerned with the protection of the environment, the provision of land for all people whatever their gender but especially for the poor and ethnic minorities, and the prevention of land speculation and the avoidance of land disputes. As Burns et al. have reported:

> The policy environment for land titling projects is becoming more complex, and a range of issues must now be addressed if a project is to pass through a Multilateral or Bilateral funding agency's approval process. These include impact on gender, impact on the environment, resettlement requirements and impact on indigenous groups (Burns et al. 1996).

Gender issues, for example, are becoming increasingly important with the international funding institutions demanding that gender equity be present both in law and in practice; this requires performance indicators to demonstrate compliance with the regulations. Concern for the poor is also growing since it is increasingly clear that to have legitimacy, newly registered property rights must reflect the social consensus of communities. Many governments and commercial organizations in developing countries have in the past tended to give formal recognition only to the property rights of the rich and politically influential; the poor in the informal sector are often excluded from benefits enjoyed by those in the formal sector of society. When the law allows the poor to formalize their property rights, the high transaction costs of doing so (and of keeping subsequent transactions formal) frequently result in economic discrimination.

The benefits of formal property expand and reinforce one another once the number of parcels and ownership rights that have been officially registered reaches a critical mass, and when costs of formal transactions are low enough not to act as a disincentive. For example, consumer credit companies lending to the poor will continue to use informal mechanisms to manage risks so long as formal property holders comprise a small portion of their client base. As the number of formal owners increases, these companies begin to change the way in which they do business. Since loans to the poor are small and for a short period, registry systems must be efficient if formal mechanisms such as mortgage registrations are to be cost effective alternatives.

The relationship of property to the environment is also receiving greater attention. Since 1988, all World Bank projects have been rated as to their potential environmental impacts. The relationship of property reform projects to the environment has been expressed in such terms as: 'improved soil conservation because of more tree crops; less intensive use of land resulting from better farming practices; and in urban areas, benefits related to health and well being, from more potential private investment in environmental infrastructure in housing, and the resulting investment by the public sector for improved drainage and sewerage' (Holstein 1996).

According to Burns et al. (1996), project managers have also had to contend with:

1. inconsistencies in development and investment priorities;
2. a lack of appreciation of the importance of focusing on policy development as compared to more tangible infrastructure development;
3. vested interests; and
4. bureaucratic inertia and ill-defined political will.

A further challenge in land administration projects is the long time frame often required for their implementation, often running over many decades. In many respects these time frames reflect the need for formal, participative, legal, and administrative processes to evolve and become sustainable. However the long time frames can also create tension amongst policy-makers anxious to be seen to achieve results.

13.2 Providing an Integrated Service

An underlying theme of this book has been the need to add value to the various collections of land related information that are in general held by governments. The processes of data integration are both technical and institutional. The traditional separation of the attributes of land into their various components may have administrative convenience but leads to duplication at the national level and a loss in the value of the information as a resource. By combining data sets, new information not apparent in either can be created; conversely, by fragmenting data sets, information and hence value is lost. Given the increasingly complex societies in which all humans live, the optimum use and sustainability of land as a resource becomes essential. The best way to achieve this is through partnerships between the traditional data managers, especially those concerned with land tenure, land taxation and finance, and land use planning and environmental control (Figure 13.1).

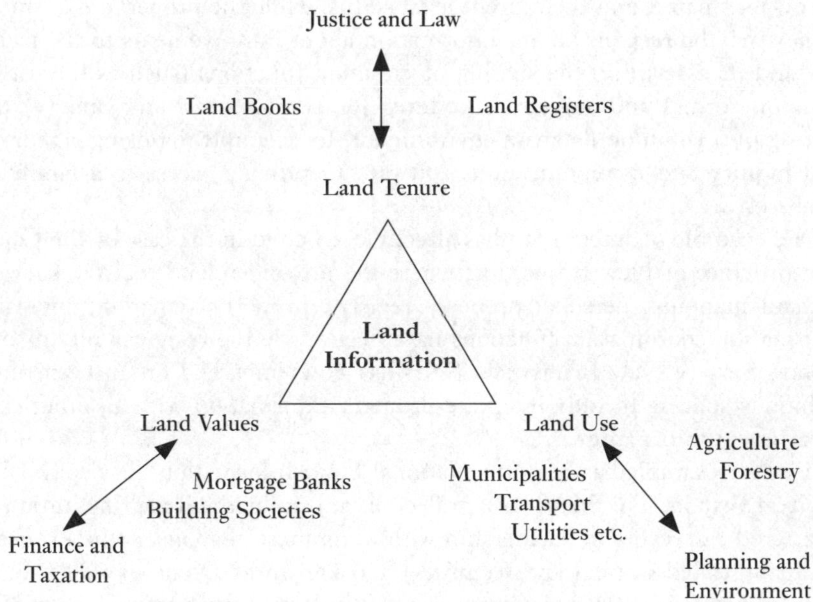

FIG. 13.1 The central role of land information

Throughout the last decade there have been moves to re-engineering the function and operation of land registries. As well, there has been growing interest in integrating land registration with other property related functions. This has in part been stimulated by the prospects of automation with the case for integration being made in terms of cost savings through:

1. economies of scale;
2. minimizing duplication;
3. opportunities to generate income by adding value to the original data;
4. more effective service to the public through 'one-stop shopping'; and
5. potential synergies, especially the sharing of service provision among the different property agencies.

The argument for greater integration is however not just commercial. It is concerned with the better management of resources to ensure that their use is sustainable for future generations. Various strategies can be adopted in order to develop partnerships between the leading participants in land administration; for example, synergies can arise through (i) a simple sharing of physical facilities by different property agencies; (ii) the sharing of staff, for example, front desk staff who receive documents and fees; (iii) the sharing of data with individual agencies given primary or 'custodial' responsibility for selected data (for instance, the land registry might be responsible for the parcel identifier, the owner's name, and the current legal status, while the property assessment agency may be responsible for information about improvements to the property and its asset); (iv) the sharing of common functions (such as the use of common forms) and similar procedures for similar functions; and (v) the sharing of a common database environment, for example, by using standardized inquiry and communications software to provide access to a family of databases.

One example of integration has already been cited in the case of the Canadian province of New Brunswick, where the provincial land registry, surveying and mapping, personal property registry, property valuation, and land information coordination functions have been made the responsibility of one agency, Service New Brunswick. SNB has now embarked on an even more ambitious scheme to fully integrate property information with all other core government information.

Another example has been the National Land Information Service (NLIS) in Great Britain. This began as a project organized by the Royal Institution of Chartered Surveyors in partnership with a commercial sponsor (the Property Company called Capital and Counties) and known as Domesday 2000. The project looked initially at the business case for setting up a network providing access to all data on the tenure, value, and use of land in Great Britain. This led to the building of a demonstrator system that was mounted on a personal computer and shown around the country. The aims of the demonstrator were to:

1. develop awareness especially among professionals working in the property sector of the potential of the technology and the growing availability of relevant data sets;
2. identify priority areas where users might most want access to data or data processing; and
3. promote the idea of establishing a national land information system.

Over 300 demonstrations were given to decision-makers and policy-formers, computer experts and cynics, and to members of the general public. The work stimulated the Land Registry in collaboration with the Ordnance Survey and the Valuation Office and one local authority (Bristol) to build a new demonstrator, this time using live data from each of their agencies. This in turn led to a pilot study in part of the city of Bristol—from which the NLIS concept was developed. This was submitted to review by a major firm of management consultants who recommended its extension across the whole country. Subsequently a version tailored to the needs of Scotland has been developed under the title SCOTLIS. Both projects will be market driven—the users must be willing to pay for the service.

Private sector involvement was of great importance as a catalyst for creating this national land information service. Since the pioneering work on the Domesday 2000 project there has been a major UK government initiative called 'government.direct'. This is concerned with improving the flow of information within government, that is between departments and ministries, and to the public, for instance, through touch screen kiosks that are being provided to help people to gain access to government information.

Some of the benefits from this major initiative in computerization will be internal to the agencies involved, benefits such as greater speed, efficiency, and economy. Other benefits will flow from the provision of a better service and greater customer satisfaction while an additional benefit should be a reduction in the number of civil servants employed by government.

The approach in the UK contrasts starkly with that elsewhere in Europe. The three leading providers of computerized access to government held property data are Sweden, Austria, and the Netherlands all of whom adopted a 'top-down' approach rather than the UK's 'bottom-up'. Sweden, with a population of under 10 million people has 20,000 terminals connected to the central database and 100,000 enquiries every day; 60 per cent or so of these enquiries are not directly concerned with property transactions. The system in Austria has some 5.5 million registered titles with 1.5 million enquiries each month. Both these systems are now self-financing.

The Austrian system was built on cooperation between those responsible for the cadastre and the land books—two separate ministries. In most countries, progress towards integration has been slow, largely because of history and rivalries between the different ministries and different professional interests. The various registries have long institutional histories and their own constituencies and professional and technical perspectives. In many cases they have made significant investments of their own in automation to address their own specific requirements. Given this, a more limited alternative has been to provide jurisdiction-wide on-line access to networks that provide query capability to the various databases and some transaction capability. These networks extend beyond real property-registry databases to those dealing with personal property, motor vehicles, health care, tax collection, etc.

13.3 Towards the Future

Electronic Title Registration

There are two logical developments that follow from computerizing land titles and ancillary land related data. The first is to automate the whole process of land transfer. Computerization allows the traditional processes of gathering, storing, retrieving, and disseminating land related data to be undertaken more quickly and with some in-built quality controls. Given access to wide area networks it is then only one small technological step further automatically and virtually instantaneously to convey the rights to a property from a vendor to a purchaser.

The second consequence of computerization is that it creates opportunities to exploit the information that is held in a land registry; new products and services can be derived by analysing and modelling the data. Thus for example assessments of property values can be automatically generated by comparison with the known market price of properties nearby, or sites can be identified that are suitable for specific types of development.

Over time, computerization will reduce the amount of office space needed to store land records and the number of people employed if the land administration system is to operate efficiently. The major initial cost lies in data conversion, a process that may take a decade or more depending on the number of records to be digitized and the acceptance or otherwise of facsimile copies rather than processed data as a final product. Facsimile copies cannot on their own provide data for automatic analysis and interpretation but can support the traditional processes of land registration.

Electronic commerce, including the opportunities for buying and selling property on the Internet, raise a number of policy issues that transcend the mere implementation of one technology rather than another. From a technical point of view, there are many processes in common with manual procedures—both are prone to error, especially human error in the data entry and data interpretation, and both are open to fraud.

Human error is particularly common when reading off a computer screen—people make an estimated 60 per cent more mistakes in detecting spelling errors when reading text off a screen rather than from a paper based printout. On the other hand, unintelligent software has difficulty in matching personal names that are not spelt identically—'Stephen' for 'Steven' or 'Alan' for 'Allan'. When responding to a request for information derived from the property owner's name, an automated system can easily ignore entries because the spelling differs from that for which the search is taking place.

All systems are open to fraud and when on-line access is granted to a database there will always be a danger that careless, criminal, or mischievous people will deliberately alter entries. There are various strategies to defeat those who attempt to break into the system, including the use of passwords and restrictions on the rights to alter data records. Most computerized registries regularly check to see if hackers can break into their system.

While computers, like humans, are not infallible, the levels of risk in managing land information are however probably less severe in a computerized system if the proper precautions have been taken. Every land registration system is prone to errors, which is why some adopt the Insurance Principle and pay compensation even though the Registrar has not been at fault. To this extent, computerization raises issues about how things are done, but from a risk management perspective changes little else.

Merely giving access to land administration data may raise issues of privacy but will not of its own affect the accuracy of the data other than in a positive way. Errors in the system are more likely to be detected and amended if those to whom they relate are allowed to see what has been recorded. If the whole process of land transfer is to be automated, however, then further issues arise especially at the time when new entries into the database have to be validated, for instance at the time of sale of a property. In many systems it is the Registrar of Titles who provides the guarantee of accuracy of all that is recorded in the registers.

The Registrar needs to know that any person purporting to sell a parcel of land is the actual owner. This cannot easily be guaranteed if the only evidence is electronic. The problem may be no different from other forms of electronic commerce (and banking provides some of the best examples of the challenges entailed). In general, however, for the individual the monetary value of the deal is likely to greatly exceed that in almost all other transactions that may take place for instance across the Internet. Using one's credit card for the purchase of a book or compact disc is substantially different in risk terms from paying for a house. Having said this, however, it should be noted that the buying and selling of real estate has quickly moved onto the web. As of 1998 there are more than 200,000 websites related to real estate sales and mortgage lending and according to one estimate (Kim 1998), more than 30 per cent of all real estate loans in the USA will be executed via the Internet by 2005.

The moment at which a transfer becomes valid also becomes important. It may be at the moment when the vendor and purchaser sign a deal or when the entry on the register is changed. If it is the former and the land registry merely keeps a record of what has occurred then this is not full automation. In manual systems the declared signature of the owner can be checked against previous documents or the integrity of the lawyers acting on behalf of the parties can be taken as a guarantee that the deal is taking place in good faith. In some systems even this is insufficient and notaries must be employed to assure that those signing deeds and other documents are the persons who they purport to be. What will be the electronic equivalent of these processes?

Of special concern in the computerized world is the issue of authentication. There is a fundamental requirement of being able to prove that a digital document and its contents have originated from where they claim to have come from, and that they have not been tampered with or altered. With the electronic transfer of land administration data, the signature could either be sent as a facsimile or else precautions similar to those involved in the prevention of hacking must be introduced. The difficulty with facsimiles is that it is so easy to scan a written signature electronically. False copies can very easily be provided. One solution is for every landowner to be given a secret 'pin number' that would be used to validate any request for transfer. This could in practice lead to confusion and chaos in its implementation since many citizens already have a variety of pin numbers.

An alternative is to insist that all transfers are validated by a notary who would have registered access to the land titles records—the Registrar would then pass on the responsibility for the accuracy of the database entries to the notary. This would perpetuate the role of the notary or, in countries where the notarial system does not already operate, cause an additional layer of bureaucracy to be introduced, thus adding to the costs. Instead, the ultimate solution is more likely to be found in the introduction of new electronic commerce technologies, such as cryptolope technology (which employs protected digital envelops), digital watermarks, and digital certificates. The latter are offered by companies such as GTE Cybertrust and serve to identify the sender and authenticate the data transmission. In the United States, the American Bar Association has prepared a draft set of guidelines for such digital certificates; the World Intellectual Property Organization is working on similar initiatives (Bacchetta et al. 1998).

Exploiting Land Administration Data

The importance of computerization lies not only in that it creates the potential for greater efficiency but also that it provides opportunities to exploit the data in a wide range of applications. These in turn can be used to generate additional income and to help the overall land administration process to be self-funding or even profit making. Up until now, very little use has been made of geographic information systems (GIS) technology in analysing land related data. Much of the use of GIS has been no more than database management, including the storage and retrieval of land records, and digital mapping including the computerized production of cadastral maps. Some agencies use GIS as an interface so that by 'clicking' on a land parcel displayed on a cadastral map, data relating to that parcel—such as its ownership, value, and use—can be displayed immediately. While such applications improve the management of the data, they do not on their own help in the management of land as a resource.

In many of the less developed countries the resources available are insufficient to permit investment in a computerized land information system. Time will remedy this. Elsewhere in the world steps are already being taken to computerize all the land records around the basic framework of land parcels. This is the first and most important and expensive step towards building a comprehensive land information service.

Once the data are stored in a computer, various types of analysis can take place. At present the understanding of the role of land and property in market economies is rudimentary. Likewise the manner in which property values move is not well modelled and a quantitative understanding of the importance of location in determining property values is limited. Research into spatial syntax is beginning to show how people perceive the space around them but the extension of this into property values has not yet provided a scientific basis for land market valuation.

Similarly the importance of understanding land tenure in planning and development control is often ignored by those concerned more with the aesthetics or the 'grand plan' than its detailed implementation. Those seeking to purchase land must all too often make separate enquiries at the land registry to find some use rights and at the local municipality to find others, an inconvenience both for those seeking sites for development as well as purchasers of properties that have already been built.

From a resource management perspective it is essential to know details about the resource that is being managed and the context within which that resource exists.

13.4 Conclusions

Land administration involves an integrated process that must operate within a framework for advancing a broad economic and social agenda, while at the same time promoting sustainable development (Figure 13.2). National politics will invariably dictate the overall agenda for development and the priorities on which the politicians may wish to focus. Land is often a key component in the successful implementation of those policies, hence the importance of a sound and comprehensive land titling and registration system. In many countries, land that is suitable for particular types of development is already in short supply. Hence the challenge is how to reconcile different proposed types of land use. In every development project there will be some winners and some losers. Without a proper understanding of land the losers may not be identified until it is too late for remedial action to be taken.

Given a market economy, the dilemma is how to treat land within the market without allowing long-term damage to result from short-term profiteering. The elements of re-engineering discussed earlier have in part been a response to the failure of many existing systems to cope with the diversity of interests and concerns that surround land and property. Land speculation, for example,

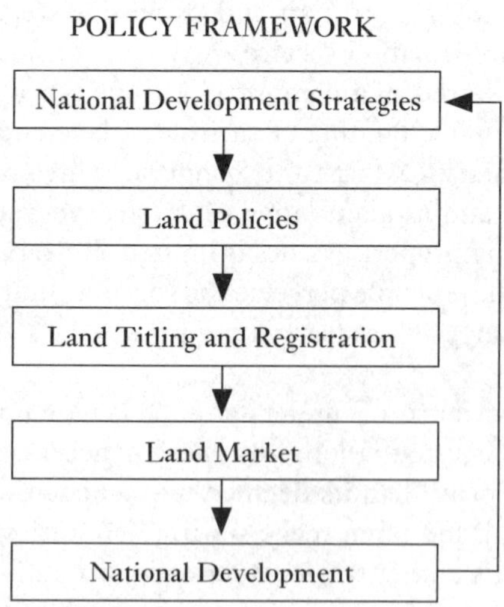

FIG. 13.2 Land policy framework

is a market process that many consider needs to be controlled if equity and justice are to prevail.

The key is to see property formalization and land titling in context, not as an end in itself but rather as a means to an end. The same may be said of the broader field described here as land administration. Underpinning both sets of processes is the management of land related information. As we said at the end of our study on land information management:

> Finally, whatever is attempted must be designed as much for the future as for the present. Maintenance is more important than initial system creation for without it, the system will become an historical monument and a folly at that. Not only must external changes be recorded within the system but also the system must itself be capable of change as the levels of sophistication both of the hardware and software and of the people operating them grow. If Third World countries are to make a quantum leap forward, if the growth of land information systems has the impact on societies that it is hoped that they will, if in fact better land information can lead to better decisions about the use of land and resources and better management of that most fundamental resource, then there is a heavy responsibility on those giving aid and assistance to get it right. The affluent can afford the occasional failure. The Third World cannot.

That principle is still true.

References

Bacchetta, Marc, et al. (1998), *Electronic Commerce and the Role of the World Trade Organization* (Geneva: WTO).

Burns, T., Eddington, R., and Nettle, K. (1996), 'Land Titling Experience in Asia', Orlando, Florida: paper presented to the International Conference on Land Tenure and Administration, November.

Holstein, L. (1996), 'Land Registration Systems,' Sydney: lecture notes prepared for the Graduate Diploma in Land Administration, University of New South Wales, August.

Kim, Jeanhee (1998), 'Home on the Net', *Money.Com.* 1/2.

Glossary of Terms

adjudication, the determination of rights in land.

adverse possession, the acquisition of land rights as a result of a period of peaceful occupation.

appraisal, estimating the market value of property.

assessment, determining the tax level for a property based upon its value relative to other properties.

boundary, either the physical objects marking the limits of a property or an imaginary line or surface marking the division between two legal estates.

cadastral parcel, a tract of land, being all or part of a legal estate.

cadastre, *juridical*: a register of ownership of parcels of land; *fiscal*: a register of properties recording their value; *multipurpose*: a register of attributes of parcels of land.

data, a raw collection of facts.

database, an organized, integrated collection of data.

digital mapping, the processes of acquisition (capture), transformation, and presentation of spatial data held in digital form.

downsizing, the reduction in numbers of staff working for an organization.

easement, a right enjoyed by the owner of land (the dominant owner) over the land of another (the servient).

fiscal value, the value of real estate property used for taxation purposes.

formalization, the process of converting de facto property rights into a formal system that is subject to statutory law.

fragmentation, the division of land into units too small for rational exploitation, usually as a result of the system of inheritance. The process may lead to a multiplicity of parcels for one owner or a multiplicity of owners of one parcel.

freehold, a free tenure, distinct from leasehold, in which the owner has the maximum rights permissible within the tenure system.

general boundary, a boundary for which the precise line on the ground has not been determined.

geodetic framework/network, a spatial framework of points whose position has been precisely determined on the surface of the earth.

Geographic Information System (GIS), a system of capturing, storing, checking, integrating, analysing, and displaying data about the earth that is spatially referenced. It is normally taken to include a spatially referenced database and appropriate applications software.

Global Positioning System (GPS), a system for fixing positions on the surface of the earth by measuring the ranges to a special set of satellites orbiting the earth.

grant, a general word to describe the transfer of property whereby rights pass from the 'grantor' to the 'grantee'.

Grundbuch, another word for Land Book (q.v.).
informal settlement, land that is occupied unofficially.
information, data transformed into a form suitable for the user.
land, the surface of the earth, the materials beneath, the air above, and all things fixed to the soil.
land administration, the processes of regulating land and property development and the use and conservation of the land, the gathering of revenues from the land through sales, leasing, and taxation, and the resolving of conflicts concerning the ownership and use of the land.
Land Book, an official record of the ownership of parcels of land.
land consolidation, the process of assembling a series of scattered land parcels into a smaller number of units.
land information system (LIS), a system for acquiring, processing, storing, and distributing information about land.
land management, the management of all aspects of land including the formation of land policies.
land parcel, the same as cadastral parcel (q.v.).
land reform, the various processes involved in altering the pattern of land tenure and land use of a specified area.
land registration, the recording of rights in land through deeds or as title.
land restitution, the process of restoring old land and property rights after the land has been taken over by the State.
land tenure, the mode of holding rights in land.
land title, the evidence of a person's rights to land.
land use, the manner in which land is used, including the nature of the vegetation upon its surface.
land value, the worth of a property, determined in a variety of ways that give rise to different estimates of value.
leasehold, land held under a lease which is a contract by which the right of exclusive possession of land is granted by a landlord (the lessor) to a tenant (the lessee) for an agreed amount of money for an agreed period of time.
local area network (LAN), a communication system that allows several processing devices that are nearby to be linked together.
lot, a cadastral parcel (q.v.).
market value, the most probable sale price of a real estate property in terms of money, assuming a competitive and open market.
mass appraisal, the process of valuing a group of real estate properties at a given date, using standard methods.
mortgage, the conveyance of a property by a debtor (called the mortgagor) to a creditor (called the mortgagee) as security for a financial loan with the condition that the property is returned when the loan is paid off by a certain date.
network (computer), a system consisting of a computer and its connected terminals and devices. The term is also used to describe two or more interconnected computer systems.
network (survey), a series of connected survey points that provide a spatial framework for an area.

orthophotograph, a composite aerial photograph from which height and tilt displacements have been removed.

orthophotomap, a photomap (q.v.) made from orthophotographs.

outsourcing, contracting out work to external bodies notably where government agencies give work to the private sector.

overriding interest, a legal interest in land that has legal force even though not recorded in the land registers.

photogrammetry, the science and art of taking accurate measurements from photographs.

photomap, a map made by printing photographs rather than using abstract conventional signs and symbols.

plot, a land parcel (q.v.).

private conveyancing, the transfer of rights in land without any public record of the transfer.

property, either the buildings associated with land or more specifically the legal rights attached to the land.

quality assurance, procedures which ensure that an organization creates products that are fit for the purposes for which they are intended.

rectification, the legal process whereby errors on a land register may be corrected.

registration of deeds, a system whereby a register of documents is maintained relating to the transfer of rights in land.

registration of title, a system whereby a register of ownership of land is maintained.

regularization, establishing formal property rights.

remote sensing, the technique of determining data about the environment from its spectral image as seen from a distance.

rental value, the value of a property in terms of the rent that may be derived from it.

restrictive covenant, an agreement whereby one landowner agrees to restrict certain ways in which the land might be used for the benefit of another.

servitude, an easement (q.v.).

spatial referencing, the association of an entity with its absolute or relative location.

stamp duty, a levy charged on the transfer of property.

strata title, title to land that is not necessarily divided horizontally, such as in high-rise buildings or for mining rights.

subdivision, the process of dividing a land parcel into smaller parcels.

systematic adjudication, the determination of rights in land on a regular and systematic basis for example within one area at one time.

title to land, see land title (q.v.).

topography, the physical features of the earth's surface.

topology, the study of properties of a geometric figure that are not dependent upon absolute position, such as connectivity.

valuation, the determination of the value of property.

Author Index

Subject Index